The Mathematics
of Inheritance Systems

David S Touretzky
Computer Science Department
Carnegie-Mellon University

The Mathematics
of Inheritance Systems

Pitman, London

Morgan Kaufmann Publishers, Inc., Los Altos, California

PITMAN PUBLISHING LIMITED
128 Long Acre, London WC2E 9AN

A Longman Group Company

© David S Touretzky 1986
First published 1986

Available in the Western Hemisphere from
MORGAN KAUFMANN PUBLISHERS, INC.,
95 First Street, Los Altos, California 94022

ISSN 0268-7526

British Library Cataloguing in Publication Data

Touretzky, David S.
  The mathematics of inheritance systems.—
  (Research notes in artificial intelligence,
  ISSN 0268-7526)
  1. Artificial intelligence—Mathematics
  I. Title   II. Series
  006.3   Q335

  ISBN 0-273-08765-7

Library of Congress Cataloging in Publication Data

Touretzky, David S.
  The mathematics of inheritance systems.

  (Research notes in artificial intelligence)
  Revision of the author's thesis (doctoral)—
Carnegie-Mellon University, 1984.
  Bibliography: p.
  1. Computer architecture.   2. Artificial intelligence.
3. Logic, Symbolic and mathematical.   I. Title.
II. Title: Inheritance systems.   III. Series.
QA76.9.A73T67   1986   006.3   86-2954
ISBN 0-934613-06-0

Reproduced and printed by photolithography
in Great Britain by Biddles Ltd, Guildford

Contents

References   **217**

Preface

This book is a revised version of my doctoral dissertation, completed at Carnegie-Mellon University in 1984. Its primary aim is to present a formal mathematical theory of a popular reasoning strategy that to date has been defended mostly by appeals to intuition: multiple inheritance with exceptions to inherited properties. Virtually all knowledge representation schemes and object-oriented programming languages include some sort of inheritance mechanism. Common examples are FRL, KRL, KLONE, NETL, Simula, Smalltalk, Flavors, LOOPS, and Ada. But the lack of a formal theory of inheritance hid some defects in existing inference algorithms. One was the incorrect treatment of networks with multiple consistent theories; another was a tendency to reason incorrectly when true but redundant statements are present. Both these problems can be eliminated once a more rigorous understanding of inheritance has been achieved.

Reasoning with exceptions is complicated because it involves operations outside of classical first order logic. The formalism I have developed to express the nonstandard inference rules that underlie inheritance bears some relation to default and nonmonotonic logics, but it includes an important hierarchical notion these other systems lack. The formalism and the definitions that follow allow us to prove theorems about the consistency, uniqueness, and constructability of inheritance theories, and lead to a formal semantics for inheritance in terms of constructable lattices of predicates.

The second major component of this thesis is the application of the inheritance theory to the analysis of a connectionist computer architecture, parallel marker propagation machines (PMPM's), of which the best-known example is Fahlman's NETL Machine. PMPM's and related connectionist schemes have aroused considerable interest as high speed inference engines for AI. The formal theory serves as a correctness specification for PMPM inheritance algorithms and allows us to show that a PMPM can reason correctly only for certain limited network topologies. However, through a technique known as "conditioning," the topology of a network can be altered to force the PMPM to produce correct results in the more general case.

The final task of this thesis is to demonstrate that the topological, network-oriented approach to reasoning as found in property inheritance systems can be successfully applied to other inference problems. We consider the problem of inheritable relations such as "bigger than" in the sentence "elephants are bigger than bread boxes," and the sorts of inferences we should be able to make from them, *e.g.*, that particular elephants are bigger than particular bread boxes, modulo known exceptions. This type of reasoning can be formalized as an extension to property inheritance, and after this is done we can return to the analysis of PMPM architectures and produce new inference algorithms, correctness specifications, and theoretical results.

D.S.T.

December, 1985

Acknowledgements

The work reported here, although applicable to inheritance systems in general, began as an attempt to answer certain questions raised by Scott Fahlman's parallel knowledge representation system, NETL. I am grateful to Scott for creating such a pleasing and stimulating intellectual puzzle, and for allowing me to happily explore its intricacies as his graduate student. One piece of the puzzle is solved now, but others remain. Scott also aided and abetted most of my other adventures as a graduate student. The combination of freedom and unhesitating support he provided was invaluable.

Jon Doyle, the second member of my committee, taught me to write mathematics. (I, however, take credit for any remaining flaws in the writing.) Over a two year period Jon and I worked together on finding the right intuition for inheritance and rigorously formalizing it. The mathematical analysis I present here would not have been possible without his guidance. Jon read countless versions of the early chapters of the thesis; his high level of enthusiasm was a wonderful antidote for occasional discouragement.

I am grateful to Dana Scott for asking some tough questions about the meaning of inheritance, which helped guide me down the path to a formal analysis. Dana also contributed to the lattice theory part of the thesis.

James Allen, the fourth member of my committee, provided insight and encouragement in several useful discussions and helped publicize the work after the defense.

I thank David Etherington of the University of British Columbia, who was kind enough to corresponded with me periodically on such topics as inheritance and default logic, for the insights he provided and for his careful reading of the final draft of the thesis.

For the revised edition published by Pitman, Sandy Koi turned almost a hundred crudely scrawled diagrams into professional quality illustrations. Readers familiar with the original dissertation will no doubt appreciate her talents as much as I now do.

During five of my six years in graduate school I was supported as a fellow of the Fannie and John Hertz Foundation. It is a pleasure to be able to acknowledge here the Foundation's generosity.

Finally, I thank my family, and my friends: Lars Ericson, Loretta Ferro, Cynthia Lamb, Al Rotella, Andi Swimmer, and Cindy Wood.

This thesis is dedicated to the Allegheny County Airport in West Mifflin, Pennsylvania, where I spent many hours as a pilot and flight instructor — a welcome respite from thinking about inheritance.

1 Inheritance Hierarchies

"This structure of concepts is formally called a hierarchy and since ancient times has been a basic structure for all Western knowledge."

— Robert M. Pirsig, *Zen and the Art of Motorcycle Maintenance*

"How anybody can get useful work done when restricted to hierarchical inheritance is beyond me; the world just doesn't work hierarchically."

— Daniel L. Weinreb, Symbolics, Inc.

1.1 Introduction

An inheritance system is a representation system founded on the hierarchical structuring of knowledge. Virtually all knowledge representation languages and object-oriented programming languages are organized around such systems. As Weinreb notes, inheritance is often extended to more complex domains than pure hierarchies (which are just tree structures), but even so, the essential idea of a hierarchical ordering of objects remains. Well-known systems with inheritance include FRL (Roberts and Goldstein, 1977), KRL (Bobrow and Winograd, 1977), SRL (Wright and Fox, 1983), KLONE (Brachman and Schmolze, 1985; Brachman, *in press*), NETL (Fahlman, 1979), Simula (Dahl, 1968), Smalltalk (Boring and Ingalls, 1982), Flavors (Weinreb, 1981), LOOPS (Bobrow and Stefik, 1981), and Ada (DoD, 1982). Until recently, despite their widespread use, inheritance systems with exceptions remained unformalized; the lack of a formal theory hid some defects in the behavior of existing systems. This thesis presents a formal mathematical theory of inheritance with exceptions and shows how to correct the flaws in existing inheritance systems. In later chapters, the formal theory of inheritance is applied to the formal analysis of a massively parallel computer architecture known as a parallel marker propagation machine, of which the most well-known example is Fahlman's NETL Machine (Fahlman, 1979). This machine has been proposed as a high speed inference engine for applications in AI.

1.2 Taxonomic hierarchies

In AI, as in other endeavors at organizing knowledge, regularity can be exploited by creating abstractions. For our purposes, an abstraction is a collection of properties shared by the members of a set. For example, if a knowledge base contains many references to gray, long-nosed, four-legged, peanut-eating jungle dwellers, we might be motivated to create an abstraction with these properties, perhaps giving it a name such as "elephant." Abstractions can also share properties, as individuals do. Elephants and sheep have some properties in common: both are warm-blooded and bear live young. One abstraction that includes both elephants and sheep is mammal. When abstractions organized by inclusion relations form a tree, the result is known as a taxonomic hierarchy, or in AI, an inheritance hierarchy. More complex organizations are possible when the tree is replaced by a general directed graph.

1.3 Advantages of hierarchical structuring

The primary advantage of hierarchical structuring is that it is an efficient method of representation. In the case of the gray, long-nosed, four-legged, et ceteras mentioned above, the naive method of representing them would list the properties of each individual separately, but after creating the elephant abstraction listing the properties the individuals have in common, we can fully describe each one simply by saying that he or she is an elephant. To efficiently represent lions and tigers and bears as well as elephants, we could create a higher class, such as mammal, to describe the properties common to all these animals.

A second advantage of hierarchical structuring, after representational compactness, is that it makes searching more efficient. Just as we can search a binary tree of alphabetized names faster than an unordered list of names, a set of assertions organized hierarchically can be searched faster than an unordered list of assertions. Often we will have more than one retrieval task in mind, with each task requiring a different organization of the hierarchy. This calls for multiple, overlapping, orthogonal groupings of properties, and is known as multiple inheritance. Multiple inheritance provides tremendous representational flexibility, but it also introduces semantic problems that do not arise in tree-structured (simple inheritance) systems. These problems will be discussed

later in the chapter.

1.4 The necessity of exceptions

Mandatory inheritance of properties is too inflexible for representing real-world knowledge (Fox, 1979). The real world contains exceptions to almost every generalization. Although most people's ideal elephant is a gray, four-legged, peanut-eating jungle dweller, there are non-gray elephants, three-legged elephants, elephants who don't eat peanuts, and elephants who don't live in jungles. If we required an abstraction to hold true for all members of a class, very few properties could be placed there. Instead, most inheritance systems allow individuals to override the properties of an abstraction that do not apply to them. For example, if we assert that Clyde is an elephant but is not gray, Clyde may inherit four-leggedness, long nosedness, jungle dwelling, and other properties from the elephant abstraction, but he will not inherit grayness.

Classes as well as individuals may have exceptional properties. It is useful to assert that mammals have four legs, but humans are mammals with only two. If we make four-leggedness a property of mammals, then lions and tigers and bears and elephants will become four-legged by inheritance. Humans won't if we state explicitly that they are two-legged. Ahab, a one-legged human, is an exception to an exception.

1.5 Two actual inheritance systems

FRL is a typical frame-based inheritance language (Roberts and Goldstein, 1977). Figure 1.1 shows how elephants, mammals, humans, and their respective numbers of legs are represented in FRL. Abstractions are encoded in FRL as *frames*, and the inclusion relation between them is called an AKO (A Kind Of) link. The properties associated with a frame are called its *slots*. In figure 1.1, each frame has a "number of legs" slot. Mammal has the value "4" in its number of legs slot, while human has the value "2". The other frames have no value in their number of legs slot.

To determine how many legs Clyde has in figure 1.1, FRL first checks the number of legs slot of the Clyde frame. Finding no value there, it proceeds up the AKO hierarchy to search the number of legs slots of higher frames. At the elephant frame the number of legs slot is also empty. At the mammal frame the value "4" is found, so FRL concludes

3

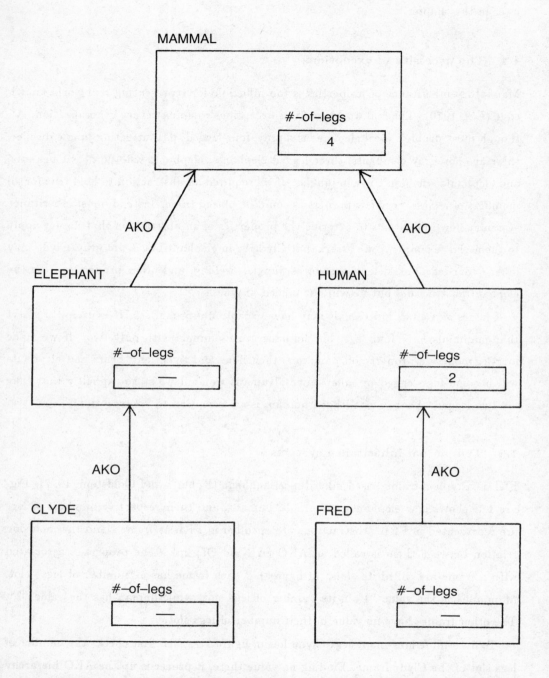

Figure 1.1: A frame-based representation of mammals, elephants, humans, and their respective numbers of legs.

that Clyde has four legs. The process we have just described is called an inheritance search algorithm. Not all systems use the same algorithm.

Suppose we ask FRL how many legs Fred has. FRL searches up the AKO hierarchy from Fred, finds the value "2" at the human frame, and the search terminates. The value "4" at mammal is never reached. That is how FRL's search algorithm accommodates exceptions.

FRL provides hooks in various places to allow a programmer to create his or her own inheritance search algorithms. I am discussing here only the most basic version of inheritance in FRL, the one provided by default.

Figure 1.2 shows one way to represent the same mammal/elephant/human example

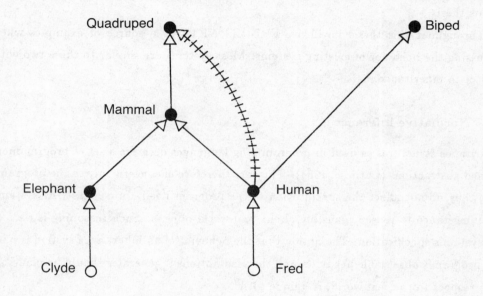

Figure 1.2: A representation of mammals, elephants, humans, and their respective numbers of legs in NETL, a semantic network system.

in NETL, a parallel semantic network language created by Scott Fahlman (Fahlman, 1979). In NETL's graphical representation, individuals such as Clyde and Fred are shown as nodes drawn as open circles, while abstractions such as elephant and mammal are drawn as solid circles. The inheritance relation, called a *VC (Virtual Copy) link, is drawn as an arrow with a closed head. Instead of an explicit representation of the "number of legs" property there is an abstraction called quadruped from which mammal inherits, and an abstraction called biped that human inherits from. The crosshatched arrow from human to quadruped is called a *CANVC link; it cancels the quadruped property for humans.

NETL does not search its knowledge base the way FRL or other serial reasoners do. NETL searches in parallel by propagating markers through the inheritance graph and drawing conclusions from where the markers land. A formal analysis of this procedure, which is known as parallel marker propagation, is given in chapter 4.

There are other ways to represent the above example in NETL, but they require additional machinery besides the basic inheritance relation, *e.g.*, relational links, or a combination of role nodes and *EQ (equality) links. Such machinery is beyond the scope of this thesis.

Throughout this thesis I will use FRL and NETL as a source of examples when discussing the behavior of existing systems. Most systems are similar to these two with respect to inheritance.

1.6   Normative inference

Inheritance structures as used in programming languages describe a set of programmer-defined abstractions (Carnese, 1984). They are therefore of concern only to the programmer; they do not affect the specification of the program itself. In contrast, AI systems use inheritance to reason plausibly about real-world objects. Such reasoning is part of the system's specification. The notion that the behavior of an inheritance system is part of a program's observable behavior, and the assumptions it generates should be plausible with respect to the real world, is unique to AI.

Much of the real-world knowledge encoded in an AI inheritance system is normative information. Normative statements are statements that are usually true, or that can be assumed to be true in the absence of contrary information. "Humans have two legs" is a normative statement. It is true of most humans. It is true of the prototypical human. But it is not universally true because there are some humans with fewer than two legs. Normative inference is important in AI because it allows a program to reason effectively when faced with incomplete knowledge about the world. If a normative reasoner knows only that Fred is human, it will assume that Fred has two legs.

Inheritance systems are efficient at producing normative inferences, but inheritance reasoning is not *necessarily* normative reasoning. An inheritance system can still represent a collection of individuals correctly if we make wholly non-normative statements about classes (*e.g.*, most mammals have six legs; most elephants are green), as long as we patch things up by putting in exceptions where necessary (Brachman, 1985). If a knowledge base consists mostly of facts about a few specialized areas, we may actually be able to make fewer exceptions by adopting a set of assumptions that are non-normative when viewed in a larger context. For example, a system that, among the mammals, reasons exclusively about people and chimpanzees, might get by with fewer exceptions if it assumed that mammals have two legs. (If there was any reason for it to know about mammals at all.)

Inheritance can at best crudely approximate normative reasoning. Consider the following statements:

Mammals have four legs.

The typical mammal has four legs.

Normal mammals have four legs.

Nearly all mammals have four legs.

Mammals may safely be assumed to have four legs.

The default number of legs for a mammal is four.

These all have similar but slightly different meanings; the differences are rather subtle. Consider the difference between "typical mammals" and "normal mammals": the latter phrase implies more latitude for deviation from the ideal than the former. But how much more? What defines the ideal? What does "normal" mean? All of the above statements would be represented exactly the same way in an inheritance system: by placing a "4" in the number of legs slot of the mammal frame. The subtleties are lost.

1.7 Multiple inheritance

If Clyde is both an elephant and a circus performer, then he should inherit properties from both those classes. But if the inheritance hierarchy is a tree he can inherit from only one superclass. Tree-structured inheritance systems force us to derive new subclasses from the natural classes that make up a domain. For instance, to efficiently represent

7

several elephants who are circus performers, we would have to create a subclass of elephant called "elephant who is a circus performer" and restate all the properties of circus performers for that class. Or else we could create a subclass of circus performer called "circus performer who is an elephant," and restate all the properties of elephants for it. Either solution is unsatisfying because it forces us to duplicate information. That we must arbitrarily choose between the two solutions, designating either elephant or circus performer as the major class, is another undesirable characteristic. These difficulties have long been recognized. Allowing a class to inherit from *multiple* superiors relieves us of the necessity of creating unneeded subclasses and duplicating information to maintain a tree-structured hierarchy.

In systems that permit multiple inheritance, the inheritance tree is replaced by a directed inheritance graph. Fahlman calls directed *acyclic* inheritance graphs "tangled hierarchies." These graphs are more difficult to compute with than trees, since the number of inference paths is (in the worst case) exponential in the number of nodes rather than linear, but massively parallel computer architectures such as NETL promise to restore the lost speed.

Some writers have proposed using taxonomic *lattices* as inheritance structures, rather than trees or directed acyclic graphs (Parker-Rhodes, 1978). An important property of lattices, but not an obviously useful one for inheritance purposes, is that every pair of elements has a least common superior (the join) and a greatest common subordinate (the meet), and these are necessarily unique. Directed acyclic graphs are posets (partially-ordered sets) rather than lattices. Posets are more general structures than lattices because they do not require unique meets and joins.

1.8   Problems with multiple inheritance

Multiple inheritance systems that admit exceptions open themselves to two semantic problems, *i.e.* there are two reasons why the meaning of their inheritance relation may become unclear or unsatisfactory. Let's start with the following example (figure 1.3):

Elephants are gray.
Royal elephants are elephants.
Royal elephants are not gray.
Clyde is a royal elephant.

8

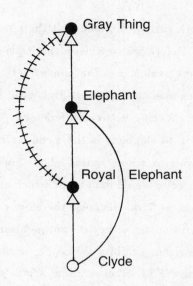

Figure 1.3: The redundant statement that Clyde is an elephant can cause problems for some inheritance reasoners.

Since Clyde is a royal elephant, and royal elephants are not gray, Clyde is not gray. On the other hand, we could argue that Clyde is a royal elephant, royal elephants are elephants, and elephants are gray, so Clyde is gray. Apparently there is a contradiction here. But intuitively we feel that Clyde is not gray, even though he is an elephant, because he is a special type of elephant: a royal elephant. FRL and NETL of course agree; their inference algorithms are designed to give subclasses precedence over their superclasses. But let us add one more statement to the initial set of axioms. We will add

Clyde is an elephant.

What is the effect of this statement? Clyde was indisputably an elephant from the original axioms, since royal elephants are elephants, so adding this new assertion ought not to affect any conclusions about him. Note, however, that when these assertions are represented in FRL or NETL, Clyde will inherit from two immediate superiors: elephant and royal elephant. In FRL, a frame inherits slot values from every frame to which it has an AKO link. We therefore find that Clyde is both gray and not gray. He is gray because he inherits from elephant, and he is not gray because he inherits from royal

9

elephant.

Not all systems work this way. Some halt their search for slot values as soon as a single value is found. Before we considered multiple inheritance, this was equivalent to choosing the "nearest" value, *i.e.* the one with the shortest inference path, which always favors subclasses over superclasses. But with multiple inheritance, path length is not a valid guide for choosing between inferences. Notice that in the example above the distance from Clyde to elephant is the same as from Clyde to royal elephant. In multiple inheritance systems whose search algorithms stop as soon as they find one value for a slot, Clyde's color would depend on which of his two immediate superiors the system happened to look at first. His color therefore cannot be predicted from just the statements in his description since they do not mention search order at all.

NETL's search algorithm exhibits the same problem. For instance, in the above example, the conclusion NETL reaches about Clyde's color depends on which of two markers the search algorithm propagates first. In more complicated networks the result would also be influenced by the *lengths* of the competing marker propagation paths, but these have little to do with subclass/superclass relations (Fahlman, Touretzky, and van Roggen, 1981).

Our first attempt at rationalizing these inheritance reasoners might be to ban redundant assertions such as "Clyde is an elephant," in order to prevent confusion. But such a move would grossly distort the semantics of the AKO or *VC link. A system that bans indisputably true statements from its knowledge base, especially when those statements are among its own inferences, is bound to appear awkward. Another reason not to ban any true statement is that statements that appear to be mutually redundant might be proposed and withdrawn for different reasons, perhaps by different contributors.

Finally, we cannot easily ban redundant statements because there *are* no truly redundant statements in a system that allows exceptions. A statement that supplies redundant information about a class as a whole may yet have an important effect on subclasses. For example:

> Elephants are gray.
> Gray things are drab.
> Elephants are drab.
> Royal elephants are elephants.

10

Royal elephants are not gray.

"Elephants are drab" appears redundant, because elephants are gray and gray things are drab. Yet if we have some other reason for believing that elephants are drab, independent of their grayness, then the seemingly redundant statement is important. Due to this statement, royal elephants are inferred to be drab even though they are not gray.

The second semantic problem arising out of multiple inheritance is that inheritance networks can now be ambiguous. Unfortunately, many existing inheritance reasoners are incapable of recognizing ambiguity. Consider the following set of statements:

Quakers are pacifists.
Republicans are not pacifists.
Nixon is a Quaker.
Nixon is a Republican.

Is Nixon a pacifist or not? According to FRL's inheritance algorithm he is both a pacifist and not a pacifist — a conclusion that is inconsistent just as Clyde's being gray and not gray was inconsistent. NETL's answer to this query is unpredictable; it depends on irrelevant features of the marker propagation algorithm used. The key feature of this example is that there is no basis for deciding conclusively either that Nixon is a pacifist or that he is not; the network is perfectly ambiguous. In logical terms we would say the network has two consistent but mutually incompatible theories. There is no reason to prefer one theory over the other because Quaker is neither a subclass nor a superclass of Republican.

1.9   The inferential distance ordering

If we desire an intuitively acceptable semantics for inheritance, we must be able to reason with redundant statements and we must not make unjustified choices in ambiguous situations. Obviously, search algorithms that determine inherited properties based on the *lengths* of competing inference paths are inadequate for this task. We must find some other ordering that allows one class to override another only if they are in the proper subset/superset relation. The solution used in this thesis is a topological relation called the inferential distance ordering. The essence of this ordering is that an individual or class A is "nearer" to class B than to class C iff A has an inference path *through* B to C.

11

(In acyclic graphs this means that B is between A and C.) Inferential distance isn't a measure of length; it's a measure of "between-ness." As a result, the ordering cannot be disturbed by redundant statements. If we adopt an inheritance search algorithm based on inferential distance instead of the traditional path length ordering, we will have no need to restrict redundant statements from the knowledge base.

Returning to a previous multiple inheritance problem, Clyde's color, we note from figure 1.3 that according to the inferential distance ordering Clyde is nearer to royal elephant than to elephant, since he has an inference path *through* royal elephant to elephant but not vice versa. An inheritance reasoner based on inferential distance would therefore conclude that Clyde is not gray.

Determining the inferential distance ordering in acyclic inheritance graphs is similar to topologically sorting the nodes: a subclass will always have a lower level number than its superclass. Figure 1.4 shows a topologically sorted DAG (directed acyclic graph)

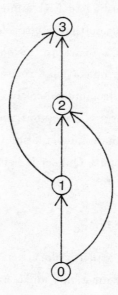

Figure 1.4: A topologically sorted DAG corresponding to figure 1.3.

corresponding to figure 1.3. We see that royal elephant, from which Clyde could inherit non-grayness, is at level 1, while elephant, from which Clyde could inherit grayness, is at level 2. If we choose among the sources of competing grayness inferences the one with the lowest level number, we would choose royal elephant over elephant and therefore conclude that Clyde is not gray.

The problem with the topological sort is that it produces a total ordering, while the ordering we intend is only partial. Inferential distance is such an ordering. Consider this example, which is similar to the Quaker/Republican problem save that the network is not symmetric (figure 1.5):

Elephants are shy.

Performers are not shy.

Circus performers are performers.

Clyde is a circus performer.

Clyde is an elephant.

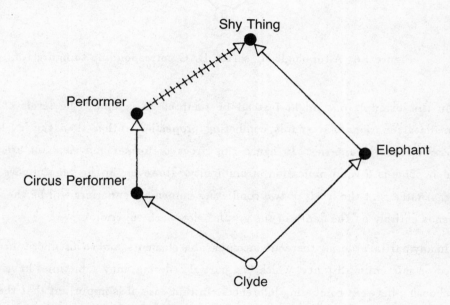

Figure 1.5: A network where topological sorting leads to unjustified conclusions.

The corresponding topologically sorted DAG is figure 1.6. Is Clyde shy or not? He could inherit "shy" via the elephant node, which is at level 1, or "not shy" via the performer node, whose level number is 2. If we simply choose the lowest number we conclude that Clyde is shy, but can such a conclusion be justified? Elephant and performer are completely unrelated classes. Since they are not hierarchically related, there is no basis for comparing the conclusions derived through them.

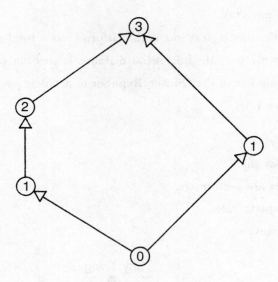

Figure 1.6: A topologically sorted DAG corresponding to figure 1.5.

In this example one might instead be tempted to compare the levels of Clyde's immediate superiors that supply conflicting properties, rather than the levels of the *sources* of those properties. In figure 1.6, circus performer and elephant are both at level 1. This is a valid indication of ambiguity. However, in the general case there is no guarantee that the levels of two conflicting immediate superiors will be the same. It depends entirely on the depth of the graph below each superior.

In any partial ordering there are incomparable elements, and so an inheritance system based on inferential distance, which is a partial ordering, may sometimes have no basis for choosing between competing inferences. In that case, it is important that the system *not* choose. Yet it should be able to reason out the consequences of all possible choices. We take the position that the network in figure 1.5 has two consistent *expansions*. An expansion corresponds to an extension in default or nonmonotonic logics (discussed below) or a theory in ordinary logic. From now on we will use the word expansion in place of the word theory due to the nonmonotonic nature of inheritance reasoning. In one expansion of figure 1.5, Clyde is shy, while in the other he is not. A correct inheritance reasoner need not actually construct all the expansions of a set of assertions, but a correct *theory* of inheritance must state what they are. It could not do so if we defined inheritance using the inference algorithms of FRL or NETL, which make no allowance for ambiguity.

1.10 A predicate logic description of inheritance

Since first order predicate calculus has a well-defined semantics (every sentence has a meaning which is derivable from its component terms and connectives according to some simple rules), one way to give a semantics to inheritance systems would be to map inheritance structures into logical sentences (Hayes, 1977). Unfortunately, the nonmonotonic nature of inheritance when exceptions are present precludes any simple axiomatization of inheritance reasoning in first order logic. Hayes, however, has argued that such reasoning *can* be expressed in first order logic by including reference to belief states (Hayes, 1979a). In this section I will merely present the standard mapping of frame-based inheritance into logical language, ignoring exceptions. Later sections consider nonmonotonic and default logics, which are better suited to representing inheritance than first order logic. In chapter 3 I will give a formal meaning to a generic inheritance system by constructing models for it using predicate lattices.

According to the usual translation of frames into logic (see, for example, (Hayes, 1979a), (Nilsson, 1980), (Hayes and Hendrix, 1981), or for less conventional translations, (Allen and Frisch, 1982) or (Attardi and Simi, 1981)), frames that represent individuals correspond to individual constants, while the slots of a frame representing an individual translate into partial functions. Thus, Clyde's color slot has a corresponding function called color, and his color would be written in logical language as color(clyde). When we put the value gray in Clyde's color slot, we are making the logical assertion color(clyde)=gray.

Frames that represent classes in FRL translate into unary predicates in logic. The elephant frame, for example, corresponds to a predicate called Elephant, and the AKO link between the Clyde and elephant frames corresponds to the logical sentence Elephant(clyde). FRL does not permit exceptions to the AKO relation, so an AKO link between two classes is a strict implication, *e.g.*, the AKO link from elephant to mammal translates into

$$(\forall x) \quad \text{Elephant}(x) \rightarrow \text{Mammal}(x).$$

15

Ignoring for the moment the fact that exceptions to slot values *are* permitted, we could represent the assertion that elephants are gray, *i.e.* the filling of the elephant frame's color slot with gray, as the sentence

$$(\forall x) \quad \text{Elephant}(x) \rightarrow \text{color}(x) = \text{gray}.$$

In the generic inheritance system defined in the next chapter we will simplify things by using predicates rather than functions to express assignments of values to slots. That is, we will assert that Clyde's color is gray by writing GrayThing(clyde), and elephants are gray by writing

$$(\forall x) \quad \text{Elephant}(x) \rightarrow \text{GrayThing}(x).$$

The advantage of this simplification is that it allows us to use the same notation for "Clyde's color is gray" as for "Clyde is an elephant." It is used in NETL for this reason. A disadvantage is that it fails to express the property that an individual can have only one color, but this is not a serious problem. Chapter 5, on inheritable relations, will present an alternative formalization of slots.

The translation of an inheritance system into logic is obviously straightforward, except for the problem of reasoning with exceptions. It is this problem that nonmonotonic logic and default logic address.

1.11 Nonmonotonic logic

A nonmonotonic logic is one where the set of conclusions derivable from a set of axioms need not increase as the axiom set is expanded. McDermott and Doyle's nonmonotonic logics, which we shall refer to as NML, are well-known formal treatments of normative reasoning (McDermott and Doyle, 1980). These logics use a modal operator M to mean "is logically consistent." A nonmonotonic inference rule such as "birds can usually fly" would be written in NML as

$$(\forall x) \quad \text{Bird}(x) \wedge M[\text{CanFly}(x)] \rightarrow \text{CanFly}(x).$$

Now let us consider an exception to the rule. Ostriches are birds but cannot fly. The recommended way to represent this information is:

$$(\forall x) \quad \text{Ostrich}(x) \rightarrow \text{Bird}(x) \wedge \sim \text{CanFly}(x).$$

16

But suppose we discover an unusual ostrich, Stretch, who *can* fly. This would be an exception to the exception. Since the original exception that ostriches can't fly was expressed as an ordinary implication rather than a nonmonotonic inference, there is no way to account for Stretch, because asserting

$$\text{Ostrich(stretch)} \land \text{CanFly(stretch)}$$

is inconsistent. Let us try a different set of axioms for birds and ostriches:

$(\forall x) \quad \text{Bird}(x) \land M[\text{CanFly}(x)] \rightarrow \text{CanFly}(x)$

$(\forall x) \quad \text{Ostrich}(x) \rightarrow \text{Bird}(x)$

$(\forall x) \quad \text{Ostrich}(x) \land M[\sim \text{CanFly}(x)] \rightarrow \sim \text{CanFly}(x)$

The problem with this set of axioms is that for an ordinary ostrich such as Henry, the assertion Ostrich(henry) generates *two* consistent expansions. In one of them Henry can't fly because he is an ostrich, but in the other, he can because he is a bird. There is nothing in NML to cause the rule about ostriches to override the less specific rule about birds. In other words, in the present system neither expansion is to be preferred over the other. Yet our intention was to assume that an ostrich cannot fly unless we have reason to believe otherwise.

In order to handle competing default rules successfully in NML we must take exceptions explicitly into account in each rule they affect. Here is a set of axioms for birds and ostriches that handles both Stretch and Henry in the desired way:

$(\forall x) \quad \text{Bird}(x) \land M[\sim \text{Ostrich}(x) \land \text{CanFly}(x)] \rightarrow \text{CanFly}(x)$

$(\forall x) \quad \text{Ostrich}(x) \rightarrow \text{Bird}(x)$

$(\forall x) \quad \text{Ostrich}(x) \land M[\sim \text{CanFly}(x)] \rightarrow \sim \text{CanFly}(x)$

Now the first rule, that birds may be presumed to fly, has been fixed by explicitly excluding it from applying to ostriches. The rule for ostriches has remained the same. Both Henry and Stretch are accommodated by these rules, but if we were to add another class of non-flying bird to the system, such as penguins, then the rule that birds can (usually) fly would again have to be rewritten to take the new exception into account. It would then look like this:

$$(\forall x) \quad \text{Bird}(x) \wedge M[\sim\text{Penguin}(x) \wedge \sim\text{Ostrich}(x) \wedge \text{CanFly}(x)] \rightarrow \text{CanFly}(x).$$

Rather than list exceptions individually as above, Doyle has suggested a mechanism for uniform defeasibility of rules by referring to each rule by name. For example, we might let "BCF" name the rule "birds can fly." The BCF rule may apply to an individual x only if the rule has not been defeated for x. If we use defeasibility of named rules to handle exceptions, we do not have to rewrite the rule for birds each time a new type of non-flying bird is encountered. However, we still must handle exceptions explicitly by giving the names of rules to be defeated. The example below shows how the BCF rule is defeated by the rule for ostriches:

$$\text{BCF}: \quad (\forall x) \quad \text{Bird}(x) \wedge M[\sim\text{DefeatedBCF}(x) \wedge \text{CanFly}(x)] \rightarrow \text{CanFly}(x)$$

$$(\forall x) \quad \text{Ostrich}(x) \rightarrow \text{Bird}(x) \wedge \text{DefeatedBCF}(x)$$

$$(\forall x) \quad \text{Ostrich}(x) \wedge M[\sim\text{CanFly}(x)] \rightarrow \sim\text{CanFly}(x)$$

The key difference between NML and inheritance systems is that, although inheritance systems are nonmonotonic, they do not require that exceptions to rules be handled explicitly. The reason is that inheritance systems assume a hierarchical structuring of knowledge, which allows exceptions at one level to *implicitly* override information from higher levels in the hierarchy. In an inheritance system, the inheritance rules recognize ostrich and penguin as subclasses of bird, so the exceptions to the rule that birds can fly need not be listed explicitly.

A hierarchical ordering of knowledge is feasible in an inheritance system because the language of discourse is limited. The inference rules of NML are much more powerful: they may include any combination of conjunctions, disjunctions, negations, nested functions, and quantifiers. There is no intuitive way to order such expressions the way classes can be ordered in an inheritance system.

From a computational standpoint, an important difference between inheritance systems and NML is that the former are designed with built-in constraints to keep them efficiently mechanizable. The greater generality of nonmonotonic logics means that they are hopelessly undecidable: in general they have no inference algorithms. Some restricted but still powerful nonmonotonic languages may have efficient inference algorithms. See the description of Etherington's work in the next section.

1.12 Default logic

Reiter's default logic is very similar to NML, except that default reasoning is expressed by rules of inference called *defaults* rather than by assertions in the logical language itself (Reiter, 1980). The description that follows is largely taken from (Etherington, 1983). Defaults are written in the form

$$\frac{\alpha(\vec{x}) : \beta(\vec{x})}{\gamma(\vec{x})}$$

where $\vec{x}$ denotes a vector of free variables and $\alpha(\vec{x})$ is called the *prerequisite* of the default, $\beta(\vec{x})$ the *justification*, and $\gamma(\vec{x})$ the *consequent*. A default says that if $\alpha(\vec{x})$ is known and $\beta(\vec{x})$ is consistent with what is known, then $\gamma(\vec{x})$ may be assumed. Reiter distinguished a natural class of defaults, known as *normal* defaults, as being of form

$$\frac{\alpha(\vec{x}) : \beta(\vec{x})}{\beta(\vec{x})}$$

In particular,

$$\frac{\mathrm{Bird}(x) : \mathrm{CanFly}(x)}{\mathrm{CanFly}(x)}$$

is a normal default rule. Reiter proved that any theory involving only normal defaults has an extension. However, as we saw in the previous section, default rules can interact with each other and generate extensions which contain undesirable conclusions. Reiter and Criscuolo discussed this possibility in (Reiter and Criscuolo, 1981) and defined a new type of rule known as a *seminormal default* whose applicability is explicitly controlled. A seminormal default is of form

$$\frac{\alpha(\vec{x}) : \beta \wedge \gamma(\vec{x})}{\beta(\vec{x})}$$

Note that all normal theories are seminormal, with a null γ. The rule

$$\frac{\mathrm{Bird}(x) : \mathrm{CanFly}(x) \wedge \sim \mathrm{Ostrich}(x) \wedge \sim \mathrm{Penguin}(x)}{\mathrm{CanFly}(x)}$$

is seminormal. Seminormal default theories are not guaranteed to have extensions. However, Etherington has shown that a subclass, known as *ordered* seminormal default theories, do have extensions (Etherington, 1983). His ordering is a dependency ordering, not the subclass/superclass ordering used by inheritance systems. By his definition, every normal default theory is ordered.

Like NML, default logic is in general undecidable. Even when we are guaranteed that an extension exists, we may have no way to construct it. In certain special cases, though, extensions are guaranteed to be constructable. Etherington gives a procedure for this.

Etherington and Reiter recently showed how multiple inheritance with exceptions could be expressed in default logic (Etherington and Reiter, 1983). But in order to control the interactions among default rules, they require the exceptions to each rule to be listed explicitly, *i.e.* they use seminormal rules. This is precisely what an inheritance system does not do. Inheritance systems use the hierarchical ordering to *implicitly* control inferences; rules are always stated in the simpler, normal form, with additional qualifications derived automatically from the hierarchical order.

In their paper, Etherington and Reiter translated a version of NETL into default logic. Their version was a combination of the one presented in (Fahlman, Touretzky, and van Roggen, 1981) and one I was experimenting with privately. The former system used a *CANCEL link to turn off *CANVC links that had been overridden by exceptions. The *CANCEL link thus appeared to block the application of a rule the way the justification part of a seminormal default would. However, *CANCEL links were purely a marker propagation device; they contributed nothing to the *meaning* of a set of assertions. The marker propagation algorithms of (Fahlman, Touretzky, and van Roggen, 1981) that made use of the *CANCEL link were later found to be incorrect, and the *CANCEL link has since been eliminated.

I believe that one of the defining properties of inheritance systems is that they use a hierarchical ordering of classes to control inference implicitly . Exceptions to a rule are never listed explicitly in these systems. Etherington has shown that default logic can model inheritance satisfactorily using seminormal defaults, but only when the implicit ordering has been made explicit. In personal communication, Etherington pointed out that inheritance rules *could* instead be expressed as normal defaults. If we choose to represent them that way, though, then afterwards we must select from the generated expansions (extensions in his terminology) those which satisfy the hierarchical requirement, *i.e.* those in which subclasses do override superclasses.

Etherington and Reiter are correct in asserting that, with some constraints, inheritance can be described as a special case of default logic. The constraints on the inheritance system are very powerful, though, and this is what makes inheritance systems worthy of study in their own right. In particular, the implicit hierarchical ordering of classes is crucial to an inheritance system, and here is where my treatment of the subject differs from theirs (Touretzky, 1984). Etherington and Reiter avoid the hierarchical issue by requiring exceptions to be explicit. In this thesis I propose an ordering — the inferential distance ordering — that allows exceptions to be made implicitly.

1.13 The meaning of normative statements

How can we decide whether a normative statement such as "typically, elephants are gray" is true in a given world? In the case of "some elephants are gray" or "all elephants are gray" there are clearly defined criteria for truth. The former statement is true if we can find at least one individual for whom the elephant and gray predicates are both true; the latter is true if we can find no individual for whom the elephant predicate is true and the gray predicate false. But there does not seem to be any comparable criterion for the truth of normative statements. Two candidates for the meaning of "typically, elephants are gray" are (1) at least k percent of them are gray, and (2) a randomly chosen elephant will appear gray with probability at least p. Both of these interpretations of "typically" have serious shortcomings. For example, neither can explain why we might want to infer "elephants are typically drab" from "elephants are typically gray" and "gray things are typically drab." That's because the conclusion is not strictly deductive, *i.e.* it does not necessarily follow from the premises. Nonetheless it would be a useful one to make in the absence of contrary information.

Interpretations based on percentages or probabilities combined with ordinary logical inference do no allow us to make reasonable assumptions about individuals. Even if we know that 99.9 percent of elephants are gray, or that randomly chosen elephants are gray with probability .999, we are not justified in inferring "Clyde is gray" from "Clyde is an elephant." Of course, nonmonotonic logic and default logic permit such inferences, but they have no need of percentages or probabilities to justify them.

In real life people often judge the "truth" of normative statements without hesitation, even when they are unable to make any percentage or probability estimates. What basis might they have for deciding that "typically, elephants are gray" is true? I believe the traditional logical notion of truth doesn't apply to normative statements. Rather, when people say that "typically, elephants are gray" is true they mean it is a useful assumption for a reasoner to make about elephants; they do not interpret it as a direct statement about probability or statistics. In a world where 99.9 percent of all elephants are gray, reasoners are almost certain to agree that "typically, elephants are gray" is a good heuristic, but in more ambiguous cases there may be legitimate disagreement among reasoners.

Neither nonmonotonic logic, default logic, nor any inheritance system offers guidance in selecting appropriate inference rules. These systems may reason with normative rules, but they could reason equally well with non-normative ones. They contain no notion of what constitutes a "good" reasoning heuristic. As users of these representations, we may prefer for reasons of efficiency or conciseness a minimal set of rules that adequately describes the domain of known individuals. But such a preference is not intrinsic to the representation, and in any event a minimal set of rules need not be normative.

I make no attempt at a theory of normative reasoning in this thesis. What I do offer is a rigorous theory of inheritance systems, which many have used as normative reasoners without fully understanding their properties. Formal analysis is now essential because the common intuitive understanding of taxonomy does not extend to systems that combine multiple inheritance with exceptions. The results reported here should give knowledge base designers and others interested in representation problems a better grasp of what multiple inheritance really means. If the reader's views on the meaning of inheritance differ from mine, he or she can still apply the techniques presented here to formalize and better understand that view.

In the next few sections I will discuss some other possibilities for accounting for normative reasoning. The chapter concludes with an outline of the remainder of the thesis.

1.14 The logic of "many" and "nearly all"

J. Altham, in *The Logic of Plurality*, tried to formalize the commonsense terms "many" and "nearly all" as nonstandard logical quantifiers (Altham, 1971). Such a logic would allow us to unambiguously formalize statements such as "nearly all elephants are gray." Altham was able to model a restricted sense of these terms using a construct called a manifold, or many-membered set. A manifold is a set of at least n distinct elements for some chosen n greater than one. The statement "many x are F," written $(Mx)Fx$ in Altham's logic, means at least n individuals are F, *i.e.* there is a manifold of individuals who are F. The statement "nearly all x are F," written $(Nx)Fx$, means fewer than n individuals are not F, which is to say that all manifolds contain an individual who is F. Note that the plural quantifier "many" resembles the existential quantifier in that it has existential import, while the penuniversal quantifier "nearly all" is like the universal in that it lacks such import. M and N, like $\exists$ and $\forall$, are interdefinable as the following identities show:

$$(Mx)Fx \quad \equiv \quad \sim(Nx)\sim Fx$$

$$(Nx)Fx \quad \equiv \quad \sim(Mx)\sim Fx$$

In domains of less than n elements, "nearly all x are F" and "nearly all x are not F" are both true. In domains of less than $2n-1$ elements, "many x are F" implies "nearly all x are F" but not the reverse. In more reasonable domains, *i.e.* ones with at least $2n-1$ elements, "nearly all x are F" implies "many x are F" but not the reverse; "nearly all x are F" and "nearly all x are not F" are mutually inconsistent; and quantifiers are ordered as follows:

$$(\forall x)Fx \rightarrow (Nx)Fx \rightarrow (Mx)Fx \rightarrow (\exists x)Fx.$$

In domains with at least $2n$ elements (enough for two disjoint manifolds) both "many x are F" and "many x are not F" can be true at once.

Altham goes on to discuss interactions between the classical and nonstandard quantifiers, and the differences between the theory of weaker systems where the domain is guaranteed only to be non-empty and that of stronger ones with at least a manifold of elements. His work, although it is an elegant mathematical exercise, still fails to capture completely the common usage of "many" and "nearly all." One example Altham gives is:

There are many severe schizophrenics, but nearly everybody is not severely schizophrenic.

This statement is inconsistent in both the weaker and stronger systems, since $(Mx)Sx$ is the opposite of $(Nx) \sim Sx$. Altham observes that the conflict seems to arise from "the casual use of these quantifiers. One tendency is to say that there are many where there are just a lot, in some vague sense, and to say that nearly all have some property when an overwhelmingly high proportion have some property.... This contradicts the interdefinability [of many and nearly all]."

There also exist valid deductions in Altham's logics that people would not accept as commonsense conclusions. For example:

All Quakers are Christians. $(\forall x) [Qx \rightarrow Cx]$

Many Quakers hate churches. $(Mx) [Qx \wedge Hx]$

So many Christians hate churches.   So $(Mx) [Cx \wedge Hx]$

The conclusion that many Christians hate churches, although valid according to the rules of the logic, is wrong. Altham explains that "the reason for this is that 'many' has *something* to do with proportions, in that the least number that can count as many of a much larger class is greater than the least number that can count as many in a much smaller class. Hence two hundred, in this case, may be enough to count as many Quakers, but too few to count as many Christians. So ... in informal argument the minimal manifold may change in size in the course of the same argument, whereas in [the logical systems presented here] the minimal manifold must remain the same size throughout any argument."

The linguistic behavior of "many" is interesting. An active sentence such as "many boys love many girls" is not equivalent in meaning to its passive form "many girls are loved by many boys," since passivization results in permuting the plural quantifiers. (Imagine a situation where each boy loves a manifold of girls, but no two boys love the same girl. Then many boys love many girls, but no girl is loved by many boys.) In contrast, "all boys love all girls" and "some boy loves some girl" are unaffected by passivization. Another interesting observation is that although the two sentences "many F are G" and "many G are F" have different meanings, they have equivalent logical representations. For example:

Many Quakers are teetotalers. $(Mx) [Qx \wedge Tx]$

Many teetotalers are Quakers. $(Mx) [Tx \wedge Qx]$

Note that we would not wish to represent "many Quakers are teetotalers" with an implication, *viz.* $(Mx)\,[Qx \to Tx]$, since that sentence is true in *any* world with a manifold of non-Quakers, even worlds in which Quakers do not exist. Our intended meaning is "there are many who are Quakers, and many who are Quakers are teetotalers," which simplifies to $(Mx)\,[Qx \wedge Tx]$. This is logically equivalent to $(Mx)\,[Tx \wedge Qx]$.

Since the representations of "many Quakers are teetotalers" and "many teetotalers are Quakers" appear to be logically identical, we can derive the latter from the former. Yet the latter statement is false. The *subject* of a sentence obviously has some bearing on the size of the minimal manifold to use in the logical representation of that sentence.

Altham solves these problems by introducing a new class of logics which he calls *attributive* that have quantifiers of varying indices. For $k > 1$, the quantifiers M^k and N^k refer to manifolds of minimum size k. So $(M^k x)Fx$ means there is a manifold of k individuals who have property F, and $(N^k x)Fx$ means there is no manifold of k individuals who do not have property F. Every predicate F has an index associated with it that defines how many individuals it takes to have "many F." Let the index of Quaker be j and that of teetotaler be k. Since there are far fewer Quakers than teetotalers, $j \ll k$. The above two sentences do not have the same representation in an attributive system because their quantifiers are different:

| Many Quakers are teetotalers. | $(M^j x)\,[Qx \wedge Tx]$ |
| Many teetotalers are Quakers. | $(M^k x)\,[Tx \wedge Qx]$ |

From $(M^k x)Fx$ we can infer $(M^j x)Fx$ iff $k \geq j$. Given that the index of teetotalers is larger than that of Quakers, from "many teetotalers are Quakers" we can infer "many Quakers are teetotalers," but not the reverse.

A problem with attributive systems, Altham points out, is that "predicates must be ordered [by assigning them indices], and this can be done only on *empirical* grounds, not from considering the *meaning* alone of the predicates." But this is not a serious objection, "for in general, to the extent that a speaker is unclear what inferences can validly be drawn from his assertions, to that extent is he unclear what his assertions actually are. Granted the attributivity of plurality-quantifiers, it *will* be unclear what inferences are licensed unless the predicates are, at least implicitly, ordered in the way suggested."

Even if we know the index of every atomic predicate, we cannot derive the indices of complex predicates. For instance, in the sentence "Many old Quakers say 'thou' instead of 'you'," the index of "old Quakers" cannot be determined simply from the indices of Quakers and old people. Thus, Altham points out, "as the indices of atomic predicates have to be informally determined, on empirical grounds, so do indices of complex predicates, and these latter have to be determined independently of the indices of the atomic predicates from which the complexes are compounded."

We can however derive some limited information about the index of a complex predicate from the indices of its components. The index of a conjunction of predicates can be no larger than the smallest index of a conjunct. Thus we at least know that the index of "old Quakers" is no larger than the index of Quaker. Conversely, the index of a disjunction of predicates can be no smaller than the largest index of a disjunct. In the case of negation, however, there may not be sufficient information available to derive any constraints on indices, because negation is really subtraction from an implicitly specified set. Unless the implicit set is made explicit, there is no way to derive the index of the result. For example, depending on linguistic context, the phrase "non-Quakers" might mean people who are not Quakers, Christians who are not Quakers, teetotalers who are not Quakers, or people attending a Quaker meeting who are not Quakers. The index of "non-Quakers" could be different in every case.

Could a logic of "many" and "nearly all" provide a semantics for inheritance systems? Unfortunately, no. The deductive nature of classical logics, even those with non-classical quantifiers, precludes their making plausible assumptions (or any other kind of assumption) in the case of incomplete information. Thus, from $(Nx)Fx$ one cannot derive Fa, *e.g.* from "nearly all elephants are gray" we can't assume that Clyde is gray just because he is an elephant. Non-monotonic logics offer a solution to this problem.

Even though Altham's logics do not provide a suitable axiomatization for inheritance systems, I feel they are worth discussing here because he has made an admirable attempt at formalizing part of commonsense reasoning and then critically analyzing the result. AI researchers would do well to emulate his example. Hayes applied this approach to representation issues in his Naive Physics Manifesto (Hayes, 1979b). Minsky and Papert's book *Perceptrons* is another good example of the benefits the formal approach can offer AI (Minsky and Papert, 1969). Next to Altham, the example closest to my own work is Attardi and Simi's description of Omega (Attardi and Simi, 1981; Attardi and Simi, 1982). My thesis extends the area of application of formal techniques to AI by offering a formal analysis of inheritance systems with exceptions as tools for representation and, as a bonus, a formal examination of parallel marker propagation machines as architectures for AI.

1.15 Frames as prototypes

A common view of frames is as descriptions of prototypes, *e.g.* the elephant frame should describe the prototypical elephant. There is some discussion of prototypes in the experimental psychology literature, but there is no formal theory of prototypes. What constitutes a good prototype? Do prototypes always exist? What about diverse sets that have no single prototype? These are some of the questions a theory of prototypes would have to address. Since we are without such a theory, viewing frames as prototypes doesn't offer much in the way of an explanation of their meaning. It may, however, suggest a useful attitude for knowledge base designers to adopt in constructing them.

The NETL system relies explicitly on the knowledge base designer's intuitive notion of a prototype, since in NETL a node that represents a class is supposed to stand for the typical instance of that class. But, like other inheritance systems, NETL says nothing about how prototypes are to be constructed. As a prototype is a psychological object, it may be that psychological investigation will contribute some useful concrete ideas to knowledge representation theory, which would certainly be an interesting development.

1.16 Reasoning about typicality

If Clyde is an elephant and typical elephants are gray, we can infer that Clyde is gray only if we assume he is a *typical* elephant. If he is an atypical elephant we are helpless. Yet he may be atypical for some totally unrelated reason, like his having only three legs. Present-day inheritance systems provide no way to assert that an individual is atypical, nor can they determine the relevance of one atypical property to some other property. For example, if the typical elephant is eight feet high and weighs two tons, and Clyde is an elephant who is only three feet high, he is atypical in a way that makes it highly unlikely that he weighs two tons. Yet he can probably still be safely assumed to be gray.

27

Collins' work on typicality reasoning (Collins, 1978) offers hope that computers may some day be able to reason effectively about atypical individuals and relevant and irrelevant evidence in support of inferences. His proposed reasoner would appear to be far more complex than any inheritance system, though.

1.17 Outline of the thesis

Chapter 2 presents a generic multiple inheritance system. We begin with ordered pairs as assertions of facts; they are of six kinds, namely: positive, negative, or "no conclusion" assertions about individuals, and positive, negative, or "no conclusion" assertions about classes. These assertions (or ordered pairs) will be visually depicted as links in an inheritance graph; they can also be expressed as sentences in nonmonotonic logic. The next step is to construct longer sequences by putting ordered pairs together according to certain rules. Formally, we will be putting assertions together to construct proofs which can be depicted visually as *paths* through the inheritance graph. When we have constructed a complete set of proofs, we will have a theory.

In constructing a theory, it will sometimes be necessary to reject a potential proof so that others can be accepted. A system based on the path length ordering always favors the shortest proof, but as we have seen, this can lead to problems. Inferential distance provides a more sophisticated basis for choosing among proofs, one which I claim better reflects our intuitions about inheritance. I will introduce definitions for constructing proofs and theories such that the inferential distance (or "between-ness") ordering is always obeyed. As a result, the two main problems with current inheritance reasoners, failure to deal with redundant statements and failure to recognize the existence of multiple theories, are eliminated.

The mathematical rigor with which I will define inheritance in chapter 2 allows many interesting properties to be proved for the resulting inheritance system. For example, we will prove that under certain very general conditions, consistent theories always exist, they are finite, and they can be constructed from the initial set of assertions by a process of sucessive approximation. We will also prove necessary and sufficient conditions for an inheritance graph to have multiple theories.

Two more lessons can be learned from the mathematical effort in this chapter, even by readers who choose not to follow the definitions and proofs in detail. The first lesson is that seemingly *ad hoc* AI systems can still be abstracted and formalized, as is traditional in mathematical work; however, to do so we may have to venture outside the familiar territory of first order logic. The second lesson is that formalization can be of practical use. Only by presenting a formal analysis, not a piece of Lisp code, can I hope to demonstrate to the reader that my proposed interpretation of inheritance is consistent and mechanizable. I will do so in chapter 2.

Chapter 3 shows how atomic predicates such as gray and elephant, represented by nodes in the inheritance graph, can be combined to yield complex predicates such as gray elephant. After introducing some definitions from elementary lattice theory, we construct a lattice of predicates in which the nodes of the inheritance graph are represented by distinguished elements determined by the inheritance graph's expansion. (For another view of inheritance systems as lattices, see (Hewitt, Attardi, and Simi, 1980).) We then investigate the sublattice of constructable elements obtainable from the basis set of distinguished elements. This allows us to develop a mathematical semantics for inheritance networks in terms of constructable sublattices.

One problem with the semantics ordinarily used with first order logic is that it is phrased purely in terms of the extensions of predicates; it says nothing about a predicate's intension. "Elephants are gray," for example, is interpreted as meaning that the extension of elephant is contained in the extension of gray thing. However, when exceptions are permitted, this interpretation of "elephants are gray" doesn't apply. Rather, "elephants are gray" is a statement about the intension of the elephant predicate, *e.g.* if x is known to be an instance of elephant then we should assume that x is gray unless we know otherwise. The predicate lattice semantics developed in this chapter allows us to represent this type of intensional information effectively.

Chapter 4 is concerned with the efficient implementation of the inference system defined in chapter 2. The lattice notation developed in chapter 3 is used to describe the states of a parallel marker propagation machine modeled after Fahlman's NETL Machine. No notion of correctness for marker propagation algorithms would be possible without an adequate specification of what inheritance should mean. Such a specification is the fruit of the analysis in preceding chapters. When inheritance systems with exceptions are given a rigorously specified semantics, we find that parallel marker propagation machines in fact do not compute expansions correctly.

In general the original assertions that generate an expansion, when represented in graph form on a marker propagation machine, do not contain enough information to use for marker propagation purposes, even when the expansion is unique. Therefore, in order to use a network for marker propagation, we must first "condition" it by adding redundant links until enough information is present at each node to guarantee correctness of the marker propagation algorithms. A program called TINA (for Topological Inheritance Architecture) is discussed that does this conditioning automatically.

This chapter contains references to the formalism developed in chapters 2 and 3, but readers who find formalism not to their liking may proceed directly to chapter 4 after skimming just the definitions in the preceding chapters. The remainder of chapter 4 discusses various conditioning algorithms and the cost of reconditioning a network when an update is made.

Chapter 5 extends the inheritance system developed in chapter 2 to cover inheritable binary relations with exceptions. One example might be the loves relation, as in "elephants love zookeepers." If Clyde is an elephant and Fred a zookeeper, an inheritance reasoner will conclude, in the absence of information to the contrary, that Clyde loves Fred. An exception to this general statement about elephants and zookeepers might be that royal elephants do not love Fred.

Since relations are inheritable, they also interact with exceptions in the class hierarchy. The key result of chapter 5 is that binary relations are a variant form of multiple inheritance. Once binary predicates are converted to unary ones by lambda abstraction, the inheritance rules for relations strongly resemble those for the class hierarchy.

Chapter 6 covers parallel marker propagation algorithms for inheritance of binary relations as described in chapter 5, including the problem of conditioning a network containing binary relations.

Chapter 7 discusses opportunities for further research in the area of inheritance systems and their formalization.

Chapter 8 presents the conclusions of the thesis. It includes an assessment of the practical and theoretical uses of the inferential distance ordering and some remarks on the suitability of parallel marker propagation machines as inference engines for AI.

2 A Theory of IS-A Inheritance

"We consider the study of a knowledge representation system as a logic system to be of fundamental importance. In this way we isolate the basic deductive mechanisms from the intricacies of specific programming languages or implementations."

— Giuseppe Attardi and Maria Simi

"What I'm getting at is that there is a problem with exceptions. It is very hard to find things that are always true."

— Marvin Minsky

2.1 A generic inheritance system

This chapter presents a generic inheritance system with exceptions. By generic I mean simple, without frills, but perfectly adequate for everyday use. Attardi and Simi, quoted in the epigraph to this chapter, state that to effectively analyze a knowledge representation system we must isolate its intended meaning from its implementation details (Attardi and Simi, 1981). I have chosen to analyze a generic system rather than a "brand name" representation language such as FRL or NETL because my generic system avoids certain idiosyncratic behaviors of the latter systems and offers an opportunity to show how the inferential distance ordering resolves the semantic problems of multiple inheritance. Another reason for not analyzing one of the brand name systems is that they are simply too large; that is, they contain many features extraneous to the inheritance machinery we wish to study. In the latter part of this chapter we will see how the inheritance components of some familiar systems, such as FRL and NETL, can be described as special cases of the generic system defined here.

2.2 The inheritance language

Our discussion of inheritance will range over several different types of objects. They are listed here, and described in detail in the following sections.

31

- Predicates and individuals.

- Tokens, which are signed predicates or individuals.

- Inheritance assertions: ordered pairs of tokens.

- Inheritance paths: sequences of tokens of length ≥ 2.

- Inheritance networks: sets of inheritance assertions.

- Inheritance theories (expansions): closed sets of inheritance paths.

The objects of discourse in an inheritance system, *i.e.* the things an inheritance system is "about," are symbols representing individuals and predicates. We refer to the latter as *real-world* predicates to distinguish them from three-valued *lattice* predicates to be defined in chapter 3. Real-world predicates are assumed to be binary.

We derive *tokens* from real-world individuals and predicates by prefixing them with a sign: one of $+$, $-$, or $\#$. Tokens are distinguished by sign as positive $(+)$, negative $(-)$, or neutral $(\#)$, respectively; they may also be distinguished by type as either predicate or individual tokens. If Clyde is an individual and elephant is a predicate, then some examples of tokens derived from these objects are: $+$clyde, which refers to the individual Clyde; $+$elephant, which refers to the class of elephants; $-$elephant, which refers to the class of non-elephants; and $\#$elephant, which refers to things not known to be either elephants or non-elephants. The symbol Π shall denote the set of all real-world individuals and predicates referred to in an inheritance system. The symbol Θ shall denote the set of tokens derivable from the objects in Π and the three signs. Formally, we write

$$\Theta = \{+, -, \#\} \times \Pi.$$

Inheritance assertions are elements of $\Theta \times \Theta$, *i.e.* they are ordered pairs of tokens. There are six types of well-formed ordered pairs; three of these contain an individual token as the first element; the other three contain a predicate token. All well-formed ordered pairs contain a predicate token as the second element. Let a be an individual and let p and q be predicates. The six types of well-formed ordered pairs and their intuitive readings are:

Ordered Pair	Intuitive Reading
$\langle +a, +p \rangle$	a is a p.
$\langle +a, -p \rangle$	a is not a p.
$\langle +a, \#p \rangle$	No conclusion whether a is a p.
$\langle +p, +q \rangle$	p's are q's.
$\langle +p, -q \rangle$	p's are not q's.
$\langle +p, \#q \rangle$	No conclusion whether p's are q's.

For example, we represent the assertion "Clyde is gray," a statement about an individual, by the ordered pair

$$\langle +clyde, +gray.thing \rangle$$

and the assertion "elephants are gray," an assertion about a class, by

$$\langle +elephant, +gray.thing \rangle.$$

The opposite assertion, "elephants are not gray," is written

$$\langle +elephant, -gray.thing \rangle.$$

Note that the first element of an inheritance assertion may be either an individual or a predicate but is always positive, while the second element must be a predicate but may be either positive, negative, or neutral. The six types of ordered pairs listed above are the only ones we will admit to the inheritance language. There are other types with reasonable interpretations, but they will be excluded for the sake of simplicity. For example, $\langle -elephant, +gray.thing \rangle$ could mean non-elephants are gray. Assertions about the complements of predicates are not normally permitted in inheritance languages. It is left as an open problem whether inheritance systems as defined here can be successfully extended to handle such assertions. One relevant observation is that the subset ordering of the complements of a set of classes is the reverse of the ordering of the classes themselves, *e.g.* the set of elephants is a subset of (and inherits properties from) the set of mammals, but the set of non-mammals is a subset of (and presumably could inherit properties from) the set of non-elephants. In the generic language defined here, since we are unable to make inheritable statements about non-mammals or non-elephants, the only way to assert a fact about non-elephants should the need arise would be to create a new token, +non.elephant, bearing no relation to the +elephant token.

33

2.3 Ordered pairs as logical sentences

Only the first two of our six types of well-formed ordered pairs correspond to sentences of first order logic, since these are the only ones that make conclusive statements about individuals. In order to give logical interpretations to all six types of ordered pairs we must choose some other logical formalism. I will use Moore's autoepistemic logic, a variant of nonmonotonic logic (Moore, 1983). In nonmonotonic logic the modal operator M means "is consistent," and the modal operator L, defined as $\sim M \sim$, means "is provable." In autoepistemic logic L means "is believed" and M is defined as $\sim L \sim$. This is a fairly technical distinction but it is important in the handling of neutral tokens, those prefixed by $\#$.

The translation given below of ordered pairs into autoepistemic logic still describes only part of the intended meaning of a collection of ordered pairs, since it omits any notion of hierarchy or inferential distance.

Ordered Pair	Autoepistemic Logic Translation
$\langle +a, +p \rangle$	$p(a)$
$\langle +a, -p \rangle$	$\sim p(a)$
$\langle +a, \#p \rangle$	$M[p(a)] \wedge M[\sim p(a)]$
$\langle +p, +q \rangle$	$p(x) \wedge M[q(x)] \rightarrow q(x)$
$\langle +p, -q \rangle$	$p(x) \wedge M[\sim q(x)] \rightarrow \sim q(x)$
$\langle +p, \#q \rangle$	$p(x) \wedge M[M[q(x)] \wedge M[\sim q(x)]] \rightarrow M[q(x)] \wedge M[\sim q(x)]$

I have chosen an ordered pair notation for inheritance assertions rather than a more conventional logical notation, such as autoepistemic logic, for the following reasons:

- The ordered pair notation treats assertions about individuals and assertions about predicates uniformly. The generic inheritance system we are defining will use the same inference rules for both types. We retain the distinction between predicates and individuals because it reflects the intended intuitive interpretations of the ordered pairs, and because such a distinction is traditional in inheritance systems. In more sophisticated versions of our generic system, especially those where quantifiers are permitted, the distinction between predicates and individuals may be important to the inference rules.

- The ordered pair notation provides a uniform, abbreviated syntax for positive, negative, and neutral statements. The logical notation above does not.

- The ordered pair notation naturally extends to sequences of length > 2, which we will use to represent inheritance paths and define the inferential distance ordering.

- The ordered pair notation expresses as a *syntactic* constraint the limitations we have set on the assertions an inheritance system may contain. For example, there is no way to represent complicated conjunctions and disjunctions of predicates in this notation; inheritance systems, unlike nonmonotonic logic, do not admit such expressions.

2.4   Inheritance graph notation

We will represent a set of ordered pairs as a finite, labeled, directed graph called an inheritance graph. The graphical representation of inheritance systems has long been popular in AI, both because it makes the organization of a set of assertions visually apparent and because inheritance reasoners are often implemented as (serial) graph algorithms. In this thesis I have a *parallel* computing architecture in mind. The architecture and a set of graph algorithms for it are presented in chapter 4. In some physical implementations of this architecture the inheritance graph corresponds directly to the connectivity of a set of hardware elements. This is another reason why it can be informative to represent sets of assertions in graphical form.

The graph notation we will use for inheritance networks is very similar to that of NETL. Individuals and predicates make up the nodes of the graph. Individual nodes are drawn as open circles, predicate nodes as solid circles. The six types of well-formed ordered pairs are represented by three types of links, called IS-A, IS-NOT-A, and NO-CONCLUSION links. When translating from ordered pairs to links we do not distinguish between pairs whose first elements are individuals and those whose first elements are predicates. That is why we need only three types of link instead of six. If we let the variables x and y range over elements of Π (the set of nodes in the inheritance graph), then the correspondence between ordered pairs and inheritance graph links is:

35

Ordered Pair	Inheritance Graph Link
$\langle +x, +y \rangle$	x IS-A y
$\langle +x, -y \rangle$	x IS-NOT-A y
$\langle +x, \#y \rangle$	x NO-CONCLUSION y

The three types of links are shown in figure 2.1. Each IS-A link in the inheritance graph is drawn as an arrow with a closed head. Each IS-NOT-A link arrow has railroad tracks, and each NO-CONCLUSION link is an arrow drawn with a wavy line.

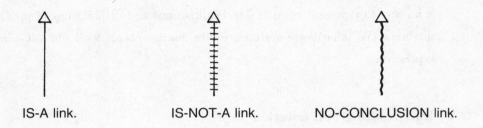

IS-A link.   IS-NOT-A link.   NO-CONCLUSION link.

Figure 2.1: The three link types.

2.5   An example: Clyde the elephant

Let us now construct a description of Clyde the elephant in our inheritance language and its corresponding graphical notation. Shown below is a collection of assertions about Clyde and his world together with their representation as ordered pairs. The graph corresponding to this set of assertions is shown in figure 2.2.

Assertion	Ordered Pair
Clyde is an elephant.	$\langle +\text{clyde}, +\text{elephant} \rangle$
Elephants are gray.	$\langle +\text{elephant}, +\text{gray.thing} \rangle$
Gray things are drab.	$\langle +\text{gray.thing}, +\text{drab.thing} \rangle$
Royal elephants are elephants.	$\langle +\text{royal.elephant}, +\text{elephant} \rangle$
Royal elephants are not gray.	$\langle +\text{royal.elephant}, -\text{gray.thing} \rangle$
Clyde is a royal elephant.	$\langle +\text{clyde}, +\text{royal.elephant} \rangle$

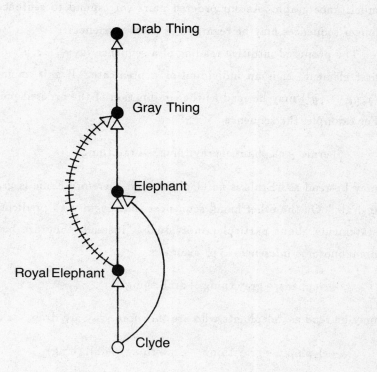

Figure 2.2: A description of Clyde.

2.6   Inheritance paths

We now consider the meanings of sequences of length greater than two. Such sequences are not part of the assertion language itself; rather, they describe paths through the inheritance graph. As our ordered pairs correspond to sentences in some logic, these longer sequences may be regarded as proof sequences.

The proposed intuitive reading of a sequence $\langle x, y_1, \ldots, y_n \rangle$ depends on whether the first element, x, is an individual or a predicate. If x is an individual, the sequence $\langle x, y_1, \ldots, y_n \rangle$ may be read as the conjunction of the ordered pairs $\{\langle x, y_i \rangle \mid 1 \le i \le n\}$. For example, the sequence

$$\langle +\text{ernie}, +\text{elephant}, +\text{gray.thing}, +\text{drab.thing} \rangle$$

may be read as "Ernie is an elephant, and therefore Ernie is gray, and therefore Ernie is drab." On the other hand, sequences that begin with predicates don't make concrete statements about particular individuals. Instead, they are best treated as chains of nonmonotonic inferences. For example:

$$\langle +\text{elephant}, +\text{gray.thing}, +\text{drab.thing} \rangle$$

may be read as "elephants who are therefore gray are drab," while

$$\langle +\text{elephant}, +\text{gray.thing}, +\text{drab.thing}, +\text{dull.thing} \rangle$$

may be read as "elephants who are therefore gray, and therefore drab, are dull." Basically, a sequence $\langle x, y_1, \ldots, y_n \rangle$ implies that x inherits, via the intermediaries y_1 through y_{n-1}, the property y_n. In contrast to the preceding example, a simple ordered pair such as $\langle +\text{elephant}, +\text{dull.thing} \rangle$ would mean that dullness was *directly* asserted to be a property of elephants.

The presence of exceptions can cause inheritance paths to be overridden. That is, certain proof sequences may be rejected because they conflict with other proofs or assertions. For example, the sequence

$$\langle +\text{royal.elephant}, +\text{elephant}, +\text{gray.thing} \rangle$$

which says that royal elephants are elephants, therefore they are gray, is overridden by the exception

$\langle +\text{royal.elephant}, -\text{gray.thing} \rangle.$

which says royal elephants are not gray. The rules that determine which sequences take precedence over others will now be explained.

2.7 The inheritance axioms

So far we have defined an inheritance language with six types of assertions, represented by ordered pairs, and described how longer sequences are to be interpreted as inheritance paths. In this section, after establishing some notational conventions, I develop a formal definition of inheritance under the inferential distance ordering. Each of the component definitions is followed by an example and/or some simple theorems to make its consequences clear. In following sections we will explore the consequences of the entire set of definitions in much greater detail.

First, recall that Π denotes the set of real-world concepts, both individuals and predicates, and Θ denotes the set of corresponding tokens (signed concepts), *i.e.* $\Theta = \{+, -, \#\} \times \Pi$. Let Σ denote the set of all sequences in Θ^* of length at least two. That is, Σ is the set of all possible sequences, both well-formed and ill-formed, of length at least two. The symbols S and Φ shall denote subsets of Σ. In particular, Φ will be used to denote expansions, or inheritance theories. Let σ denote an element of Σ, and let $\Gamma \subseteq \Theta \times \Theta \subseteq \Sigma$ denote a set of well-formed ordered pairs, or inheritance assertions.

In the definitions that follow, variables such as x and y shall range over tokens, *i.e. signed* concepts. We will use a prime notation to refer to tokens that match except for sign. That is, if x stands for $+a$, then x' can mean either $-a$ or $\#a$; if x stands for $-a$, then x' can mean either $+a$ or $\#a$, and so on.

Definition 2.1 The *conclusion set* of a set of sequences Φ, written $C(\Phi)$, is the set of all pairs $\langle x, y \rangle$ such that a sequence $\langle x, \ldots, y \rangle$ appears in Φ.

$C(\Phi)$ can be viewed as the set of facts implied by the sequences of Φ. Queries such as "is Clyde an elephant" or "are there any gray elephants" are answerable by examining $C(\Phi)$. Example: If Φ contains the sequence $\langle +\text{clyde}, +\text{royal.elephant}, +\text{elephant} \rangle$ then its conclusion set $C(\Phi)$ contains the sequence $\langle +\text{clyde}, +\text{elephant} \rangle$. The definition of $C(\Phi)$ need not include the intermediate elements of a sequence because, as we shall see, it will only be important to look at conclusions of sets that are closed with respect to contiguous subsequences.

Definition 2.2 A set of sequences Φ *contradicts* the sequence $\langle x_1, \ldots, x_n \rangle$ iff $\langle x_1, x_i' \rangle \in C(\Phi)$ for some i, $1 \leq i \leq n$.

If Φ contains a set of proof sequences, we can use the notion of contradiction to prevent any new sequence from appearing in Φ that would conflict with what is already in Φ. Example: the sequences $\langle +\text{clyde}, +\text{gray.thing} \rangle$ and $\langle +\text{clyde}, -\text{gray.thing} \rangle$ are mutually contradictory; if Φ contains one it will contradict the other.

Besides contradiction, we shall have one other way of preventing a sequence from appearing in Φ. This is a relation called *preclusion*. Φ would preclude the sequence

$\langle +\text{clyde}, +\text{royal.elephant}, +\text{elephant}, +\text{gray.thing} \rangle$

if Φ contained the sequence

$\langle +\text{royal.elephant}, -\text{gray.thing} \rangle$

In other words, since royal elephant is a subclass of elephant and royal elephants aren't gray, the argument that Clyde is gray by virtue of being an elephant is precluded by his being a royal elephant. Preclusion is what allows subclasses to override what they inherit from their superclasses.

Before giving the formal definition of preclusion we must define the *intermediaries* of a sequence, which are the subclasses that might preclude that sequence. The intermediaries to a sequence such as

$\langle +\text{clyde}, +\text{royal.elephant}, +\text{elephant}, +\text{gray.thing} \rangle$

include all the tokens that appear in the sequence. Thus, +royal.elephant is an intermediary to the above. But in addition, certain tokens which are located in the inheritance hierarchy between two elements of the sequence should be considered intermediaries to it. For example, the network in figure 2.2 which contains a redundant IS-A link can generate the sequence

$\langle +\text{clyde}, +\text{elephant}, +\text{gray.thing} \rangle$.

40

This sequence must be precluded from Φ if we are to have a sensible theory of inheritance that is not led astray by redundant links. Since royal elephant is a subclass of elephant and since Clyde is a royal elephant, +royal.elephant should be considered an intermediary to this sequence even though it doesn't appear in it.

Definition 2.3 A token y is an *intermediary* in Φ to a sequence $\langle x_1, \ldots, x_n \rangle$ iff either $y = x_i$ for some i, or Φ contains a sequence $\langle x_1, \ldots, x_i, y_1, \ldots, y_m, x_{i+1} \rangle$ where $y = y_j$ for some j, $1 \le j \le m$ and $1 \le i < n$. See figure 2.3.

The reader will no doubt have noticed that according to this definition of intermediaries it is not sufficient for y to appear between x_i and x_{i+1} in the IS-A hierarchy for it to be considered an intermediary to $\langle x_1, \ldots, x_n \rangle$ in Φ. Rather, the entire proof sequence $\langle x_1, \ldots, x_i, \ldots, y, \ldots, x_{i+1} \rangle$ must appear in Φ. The reason for this more stringent requirement has to do with the effects of multiple cancellations (exceptions to exceptions) in the IS-A hierarchy.

Definition 2.4 Φ *precludes* a sequence $\sigma = \langle x_1, \ldots, x_n \rangle$ iff Φ contains a sequence $\langle y, x_n' \rangle$ where y is an intermediary to σ in Φ.

Referring to figure 2.2, Φ precludes

$$\sigma = \langle +\text{clyde}, +\text{elephant}, +\text{gray.thing} \rangle$$

if it contains

$$\langle +\text{clyde}, +\text{royal.elephant}, +\text{elephant} \rangle$$

and

$$\langle +\text{royal.elephant}, -\text{gray.thing} \rangle.$$

Here +royal.elephant is the intermediary to σ that causes Φ to preclude σ.

Contradiction and preclusion are overlapping relations. For example, if Φ contains the three sequences

$$\langle +\text{royal.elephant}, +\text{elephant} \rangle$$

$$\langle +\text{elephant}, +\text{gray.thing} \rangle$$

41

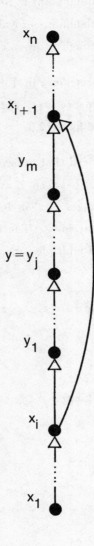

Figure 2.3: y is an intermediary in Φ to the sequence $\langle x_1, \ldots, x_n \rangle$ because Φ contains $\langle x_1, \ldots, x_i, y_1, \ldots, y_m, x_{i+1} \rangle$ and $y = y_j$ for some j, $1 \leq j \leq m$, $1 \leq i < n$.

$$\langle +\text{royal.elephant}, -\text{gray.thing}\rangle$$

then Φ both contradicts and precludes

$$\langle +\text{royal.elephant}, +\text{elephant}, +\text{gray.thing}\rangle.$$

Theorem 2.1 If a set $\Phi \subseteq \Sigma$ contradicts (precludes) a sequence σ, then every superset of Φ contradicts (precludes) σ.

Proof The conclusion set of every superset of Φ is a superset of $C(\Phi)$. Thus, if Φ contradicts σ every superset of Φ contradicts σ. If Φ precludes σ, then Φ contains a sequence $\langle y, x'_n \rangle$ where y is an intermediary to σ in Φ. If this relationship holds in Φ, it holds in every superset of Φ. ∎

Definition 2.5 A sequence $\sigma = \langle x_1, \ldots, x_n \rangle$ is *inheritable* in Φ iff $n > 2$, Φ contains both $\langle x_1, \ldots, x_{n-1} \rangle$ and $\langle x_2, \ldots, x_n \rangle$, and Φ neither contradicts nor precludes σ.

This definition, like that of intermediaries, may seem surprising at first. Why not define inheritability as the concatenation of sequences $\langle x_1, \ldots, x_i \rangle$ and $\langle x_i, \ldots, x_n \rangle$ instead of requiring an $n - 2$ overlap between the two sequences? Again, the answer has to do with the effects of exceptions on inheritance theories. For more details, refer to section 2.16 on alternative definitions for inheritability.

Note that σ's inheritability in Φ is independent of whether σ is contained in Φ. Example: If Φ consists of the two sequences

$$\langle +\text{clyde}, +\text{elephant}, +\text{gray.thing}\rangle$$

and

$$\langle +\text{elephant}, +\text{gray.thing}, +\text{drab.thing}\rangle$$

then

$$\langle +\text{clyde}, +\text{elephant}, +\text{gray.thing}, +\text{drab.thing}\rangle$$

is inheritable in Φ.

Definition 2.6 Φ is *closed under inheritance* iff Φ contains every sequence inheritable in Φ.

43

Definition 2.7 Φ is an *expansion* of a set $S \subseteq \Sigma$ iff $\Phi \supseteq S$ and Φ is closed under inheritance.

Definition 2.8 Φ is *grounded* in a set $S \subseteq \Sigma$ iff every sequence in $\Phi - S$ is inheritable in Φ.

Groundedness is interesting only for sets that are closed under inheritance. Although one can construct sets which are not closed under inheritance but are grounded in some inheritance graph Γ, such sets might have no expansions grounded in Γ. For example, see figure 2.4. The set of ordered pairs corresponding to this network is

$$\Gamma = \{\langle +a, +b\rangle, \langle +b, +c\rangle, \langle +c, +d\rangle, \langle +a, +c\rangle, \langle +b, -d\rangle\}.$$

Now let

$$S = \Gamma \cup \{\langle +a, +c, +d\rangle\}.$$

Note that $\langle +a, +c, +d\rangle$ is inheritable in S. (Since $\langle +a, +b, +c\rangle \notin S$, b is not an intermediary in S to $\langle +a, +c, +d\rangle$, so S does not preclude $\langle +a, +c, +d\rangle$.) Therefore S is grounded in Γ, but not closed under inheritance. Now let Φ be an expansion of the grounded set S. There is only one expansion of S grounded in S, namely $\Phi = S \cup \{\langle +a, +b, +c\rangle\}$. But Φ precludes $\langle +a, +c, +d\rangle$, since b *is* an intermediary to $\langle +a, +c, +d\rangle$ in Φ. Thus $\langle +a, +c, +d\rangle$ is not inheritable in Φ even though it is an element of Φ. So although Φ is an expansion of Γ, it is not grounded in Γ. S has no expansions grounded in Γ; the fact that S itself is grounded in Γ is of no interest.

Definition 2.9 Φ is a *grounded expansion* of a set $S \subseteq \Sigma$ iff Φ is an expansion of S and Φ is grounded in S.

Although the three preceding definitions (of expansions, groundedness, and grounded expansions) apply to any set of sequences $S \subseteq \Sigma$, most of the time we will be concerned with grounded expansions of Γ, where Γ is a set of ordered pairs denoting an inheritance network.

Theorem 2.2 If Φ is a grounded expansion of Γ, then $\langle x, y\rangle \in \Phi$ iff $\langle x, y\rangle \in \Gamma$. In other words, the only ordered pairs in a grounded expansion are those in the inheritance graph.

44

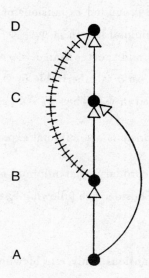

Figure 2.4: Groundedness is interesting only for sets closed under inheritance.

Proof If $\langle x, y \rangle \in \Gamma$ then $\langle x, y \rangle \in \Phi$ because $\Phi \supseteq \Gamma$. If $\langle x, y \rangle \in \Phi$ then $\langle x, y \rangle \in \Gamma$ because Φ is grounded and sequences of length two are not inheritable in Φ. ∎

Theorem 2.3 If a grounded expansion of Γ contains the sequence $\langle x_1, \ldots, x_n \rangle$, then it also contains all contiguous subsequences $\langle x_i, \ldots, x_j \rangle$ for $1 \leq i < j \leq n$.

Proof For $n = 2$ the statement is vacuously true. Suppose it is true for all sequences of length less than $n > 2$. Since the expansion is grounded in the set of ordered pairs Γ, by the previous theorem it follows that the sequence $\langle x_1, \ldots, x_n \rangle$ must be inheritable in it. Therefore the expansion contains $\langle x_1, \ldots, x_{n-1} \rangle$ and $\langle x_2, \ldots, x_n \rangle$. Thus by the inductive hypothesis, the expansion contains $\langle x_i, \ldots, x_j \rangle$ for $1 \leq i < j \leq n$. ∎

Corollary 2.1 If Φ is a grounded expansion of Γ and $\langle x_1, \ldots, x_n \rangle \in \Phi$, then $\langle x_i, x_j \rangle \in C(\Phi)$ for $1 \leq i < j \leq n$.

This corollary is what allows us to define $C(\Phi)$ in terms of the first and last elements of a sequence $\langle x, \ldots, y \rangle$ in Φ, ignoring the intermediate elements. The corollary shows that it is safe to do so when Φ is a grounded expansion.

Theorem 2.4 Let Φ_a and Φ_b be two grounded expansions of Γ. Then $\Phi_a \subseteq \Phi_b$ iff $\Phi_a = \Phi_b$.

Proof Let Φ_a and Φ_b be two grounded expansions of Γ such that $\Phi_a \subseteq \Phi_b$. Let $\sigma = \langle x_1, \ldots, x_n \rangle$ be a sequence of minimal length in $\Phi_b - \Phi_a$. $\langle x_1, \ldots, x_{n-1} \rangle$ and $\langle x_2, \ldots, x_n \rangle$ are in Φ_a because they are in Φ_b and shorter than σ. Φ_a neither contradicts nor precludes σ, since if it did Φ_b would too, so σ is inheritable in Φ_a. But since Φ_a is closed under inheritance, σ must be contained in Φ_a. Thus $\Phi_b - \Phi_a$ is empty. ∎

Corollary 2.2 Grounded expansions are minimal expansions.

The converse of the above corollary, that minimal expansions are grounded expansions, is not necessarily true. Consider the following example (due to Jon Doyle):

$$\Gamma = \{\langle +a, +b \rangle, \langle +b, +c \rangle\} \, .$$

This Γ has three minimal expansions, only one of which is grounded. The grounded expansion is

$$\Phi_1 = \Gamma \cup \{\langle +a, +b, +c \rangle\} \, .$$

The two non-grounded but still minimal expansions are

$$\Phi_2 = \Gamma \cup \{\langle +a, -c \rangle\} \, ,$$

$$\Phi_3 = \Gamma \cup \{\langle +a, \#c \rangle\} \, .$$

Theorem 2.5 If $\langle x_1, \ldots, x_n \rangle$ is an element of a grounded expansion of Γ, then x_1 through x_{n-1} are positive.

Proof Γ is a set of well-formed ordered pairs. Therefore x_1 is positive for every $\langle x_1, x_2 \rangle \in \Gamma$. Now suppose the claim is true for all sequences of length less than n, for some $n > 2$. If $\langle x_1, \ldots, x_n \rangle$ is in the expansion it must be inheritable in that expansion, so the expansion also contains $\langle x_1, \ldots, x_{n-1} \rangle$ and $\langle x_2, \ldots, x_n \rangle$. Then by the inductive hypothesis, x_1 through x_{n-1} are positive. ∎

An expansion corresponds to a *theory* in ordinary logic, or an *extension* in Reiter's default logic. Given an inheritance graph described as a set of ordered pairs Γ, the above definitions and theorems describe the structure of the expansions of that graph. The expansions contain maximal sets of permissible inferences that may be drawn from the graph.

2.8 Clyde the elephant, revisited

Let Γ denote the set of inheritance assertions describing Clyde the elephant and his world, shown in graphical form in figure 2.2. Γ has a single grounded expansion Φ, consisting of the elements of Γ plus the following additional sequences:

⟨+elephant, +gray.thing, +drab.thing⟩

⟨+clyde, +royal.elephant, +elephant⟩

⟨+clyde, +royal.elephant, −gray.thing⟩

$C(\Phi)$, the set of conclusions drawn from the expansion Φ, consists of the elements of Γ plus

⟨+elephant, +drab.thing⟩

⟨+clyde, −gray.thing⟩

Thus we conclude that elephants are drab, and Clyde is not gray. Under our definition of inheritability, Φ precludes

⟨+clyde, +elephant, +gray.thing⟩

because it contains

⟨+clyde, +royal.elephant, +elephant⟩

and

⟨+royal.elephant, −gray.thing⟩.

Note also that since Clyde is not gray, we derive no conclusion about his drabness, despite the potentially misleading path in figure 2.2 from Clyde to elephant to gray thing to drab thing.

2.9 Independence of groundedness and closure

Let Γ be the set $\{\langle +a, +b \rangle, \langle +b, +c \rangle\}$. We demonstrate that groundedness and closure are independent properties by exhibiting sets Φ that are supersets of Γ and are grounded or not grounded and independently closed or not closed, as follows:

Grounded and not closed: Let $\Phi = \Gamma$.

Grounded and closed: Let $\Phi = \Gamma \cup \{\langle +a, +b, +c \rangle\}$.

Not grounded and not closed: Let $\Phi = \Gamma \cup \{\langle +d, +e \rangle\}$.

Not grounded, but closed: Let $\Phi = \Gamma \cup \{\langle +a, +b, +c \rangle, \langle +d, +e \rangle\}$.

2.10 Ordering relations

In this section we use the graph Γ to define some ordering relations on nodes, tokens, and sequences. These relations, written $\prec$ and $\ll$, will appear in several proofs in later sections of this chapter.

Definition 2.10 Let x and y be elements of Π. Then $x \prec y$ (read x is *below* y with respect to Γ) iff either $\langle +x, +y \rangle \in \Gamma$, or for some $z \in \Pi$, $x \prec z$ and $z \prec y$.

We write $\prec$ instead of $\prec_\Gamma$ because the Γ can safely be left implicit. In figure 2.2, the graph Γ contains $\langle +\text{royal.elephant}, +\text{elephant} \rangle$, so royal.elephant $\prec$ elephant. Γ also contains $\langle +\text{elephant}, +\text{gray.thing} \rangle$, so elephant $\prec$ gray.thing. Therefore, by transitivity, royal.elephant $\prec$ gray.thing. Note that $\prec$ is strictly transitive while IS-A need not be when exceptions are allowed.

Corollary 2.3 The $\prec$ relation is a partial ordering iff the IS-A subgraph of Γ is acyclic.

We will say that Γ is "*i.a.*" when we mean it is IS-A acyclic. All inheritance systems with which I am familiar require the inheritance graph to be *i.a.* An obvious reason for such a restriction is ease of implementation, since in an acyclic system the programmer is relieved of the responsibility for preventing infinite loops in the search algorithms. Another reason might be the lack of a clear semantics for cyclic inheritance structures, especially when exceptions are present. In the generic system defined in this chapter there is no requirement that the inheritance graph be acyclic. However, some of the theorems depend on $\prec$ being a partial ordering, which means Γ must be *i.a.* Two theorems depend on the entire graph being acyclic. Can one extend the proofs to cover the general case, or find counterexamples that would legitimize an acyclicity requirement?

Definition 2.11 Let x and y be elements of Π. Then $x \ll y$ (read x *precedes* y with respect to Γ) iff at least one of $\langle +x, +y \rangle$, $\langle +x, -y \rangle$, or $\langle +x, \#y \rangle$ is in Γ, or for some $z \in \Pi$, $x \ll z$ and $z \ll y$.

We use $\ll$ as an abbreviation for $\ll_\Gamma$, with the Γ left implicit. Basically, $\ll$ is the "less than" relation under all links in Γ, while $\prec$ is the "less than" relation under IS-A links alone. The $\prec$ relation is a subset of $\ll$, and the IS-A relation (defined explicitly by $C(\Phi)$, where Φ is a grounded expansion of Γ) is a subset of $\prec$ when exceptions are allowed. The $\prec$ and $\ll$ relations are transitive, while IS-A is not.

Corollary 2.4 The $\ll$ relation is a partial ordering iff the inheritance graph is totally acyclic.

Definition 2.12 Let x and y be tokens in Θ. Then $x \prec y$ (or $x \ll y$) iff the $\prec$ or $\ll$ relation, respectively, holds for the corresponding nodes in Π.

Definition 2.13 x is a *minimal* node (token) in Γ iff there is no node (token) y such that $y \prec x$.

Theorem 2.6 If $\langle x_1, \ldots, x_n \rangle$ is an element of a grounded expansion Φ of Γ, then $x_i \prec x_j$ for $1 \leq i < j < n$.

Proof If $\langle x_1, \ldots, x_n \rangle$ is an element of a grounded expansion Φ of Γ, then x_1 through x_{n-1} are positive, and Γ contains $\langle x_i, x_{i+1} \rangle$ for $1 \leq i < n$. Thus $x_i \prec x_{i+1}$ with respect to Γ for $1 \leq i < n - 1$. By transitivity, $x_i \prec x_j$ with respect to Γ for $1 \leq i < j < n$. ∎

Now while the set of nodes Π (and hence the set of tokens Θ) is finite, the set of all sequences over Θ is infinite, so it is natural to wonder whether finite graphs Γ can produce infinite theories (grounded expansions). The next theorem assures us that this cannot happen if Γ is i.a.

Theorem 2.7 Every grounded expansion of an IS-A acyclic Γ is finite.

Proof We will show that no token appears twice in any sequence. Since Γ is i.a., $\prec$ is a partial ordering. For every sequence $\langle x_1, \ldots, x_n \rangle$ in a grounded expansion of Γ, $x_i \prec x_j$ for $1 \le i < j < n$. If x_n is positive, then $x_{n-1} \prec x_n$ as well. Therefore no positive token appears twice in any sequence. If x_n is negative or neutral, it is the only such token in the sequence. Under these conditions, since Θ is finite, there can be only a finite number of inheritable sequences. Therefore, the expansion is finite. ∎

Finally, we define a $\prec$ ordering on sequences in terms of the ordering of their penultimate tokens:

Definition 2.14 $\langle x_1, \ldots, x_n \rangle \prec \langle y_1, \ldots, y_m \rangle$ with respect to Γ iff $x_{n-1} \prec y_{m-1}$.

Corollary 2.5 If Γ is IS-A acyclic then the $\prec$ relation on sequences is a partial ordering.

Corollary 2.6 Let $\langle x_1, \ldots, x_n \rangle$ be a sequence in a grounded expansion of an IS-A acyclic Γ. Then $\langle x_1, \ldots, x_i \rangle \prec \langle x_1, \ldots, x_{i+1} \rangle$ for $1 < i < n - 1$.

2.11   Consistency

Recall the definition of *contradiction* given earlier in the chapter: that a set of sequences Φ *contradicts* the sequence $\langle x_1, \ldots, x_n \rangle$ iff $\langle x_1, x_i' \rangle \in C(\Phi)$ for some i, $1 \le i \le n$. We define consistency as an absence of contradiction. Formally:

Definition 2.15 A set of sequences is *consistent* iff it contradicts none of its elements.

Theorem 2.8 A grounded expansion Φ of Γ is consistent iff Γ is.

Proof If Γ is inconsistent then Φ is inconsistent, because $\Phi \supseteq \Gamma$. Conversely, if Φ is inconsistent, it contains sequences $\langle x_1, \ldots, x_n \rangle$ and $\langle y_1, \ldots, y_m \rangle$ such that $x_1 = y_1$ and $y_m = x_i'$ for some i, $1 \le i \le n$. Since Φ is grounded, Φ also contains $\langle x_1, \ldots, x_i \rangle$. Φ contradicts both $\langle x_1, \ldots, x_i \rangle$ and $\langle y_1, \ldots, y_m \rangle$, which are mutually contradictory as well, so neither is inheritable in Φ. But every element of a grounded expansion is either inheritable in that expansion or is an element of Γ, so Γ must contain them both. Γ is therefore inconsistent. ∎

Corollary 2.7 If Φ is a grounded expansion of Γ, then $C(\Phi)$ is consistent iff Γ is.

50

A grounded expansion of Γ can be inconsistent only if Γ contains a pair of sequences $\langle x, y \rangle$, $\langle x, y' \rangle$, or a single sequence $\langle x, x' \rangle$. An inconsistent $\Gamma = \{\langle +p, +q \rangle, \langle +p, -q \rangle\}$ is shown in graph form in figure 2.5. In contrast, the two graphs in figure 2.6 are both consistent, and so each has a consistent grounded expansion Φ. In 2.6a we have

$$\Phi = \Gamma = \{\langle +a, +b \rangle, \langle +b, +c \rangle, \langle +a, -c \rangle\}$$

while in 2.6b we have

$$\Phi = \Gamma = \{\langle +a, +b \rangle, \langle +b, -c \rangle, \langle +a, +c \rangle\}.$$

The graph in figure 2.7 has two consistent grounded expansions Φ_1 and Φ_2, but the union of the two is inconsistent because Φ_1 contains $\langle +a, +b, +d \rangle$ while Φ_2 contains $\langle +a, +c, -d \rangle$.

Figure 2.5: An inconsistent network.

Our notion of consistency differs from that of nonmonotonic or autoepistemic logic. In logical systems, a set of sentences is inconsistent iff it entails a pair of sentences of form α and $\sim \alpha$. In our inheritance system this corresponds to sequences such as $\langle +a, +p \rangle$ and $\langle +a, -p \rangle$, where a is an individual and p a predicate. But in addition any network containing both $\langle +p, +q \rangle$ and $\langle +p, -q \rangle$ is inconsistent according to our definition, although the logical translation of this pair of assertions, repeated below, is not of form α and $\sim \alpha$.

$\langle +p, +q \rangle \qquad p(x) \wedge M[q(x)] \rightarrow q(x)$

$\langle +p, -q \rangle \qquad p(x) \wedge M[\sim q(x)] \rightarrow \sim q(x)$

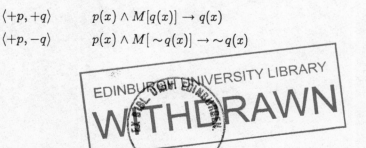
51

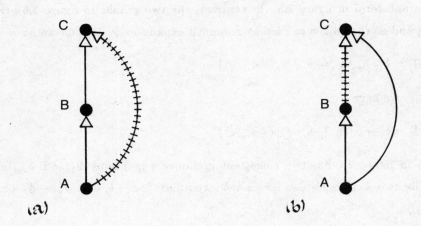

(a)   (b)

Figure 2.6: Two consistent networks, each containing an exception.

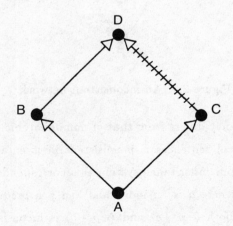

Figure 2.7: A consistent network with two grounded expansions. Their union is inconsistent.

In autoepistemic logic these two assertions may coexist with no logical inconsistency. A system containing both of them simply has two expansions, but neither one would be inconsistent. To see why this example is judged inconsistent in an inheritance system, consider the intuitive meanings we assign to the two assertions when viewed as inheritance rules. One means that p's are q's while the other means p's are not q's. When stated in English as normative inference rules the inconsistency becomes obvious; the two rules can't both be normative. It is this sense of inconsistency, not the syntactic one found in logic, that we wish to adopt for our inheritance system.

One way to express this notion of inconsistency syntactically is to define a "typical" individual t_p for each predicate $p \in \Pi$, so that $p(t_p)$ is true by definition, and then say that the correct logical translation of a sequence $\langle +p, +q \rangle$ is really two sentences, namely

$$p(x) \wedge M[q(x)] \rightarrow q(x)$$

and

$$q(t_p).$$

Similarly, we can make the translation of $\langle +p, -q \rangle$ be

$$p(x) \wedge M[\sim q(x)] \rightarrow \sim q(x)$$

and

$$\sim q(t_p).$$

If we do this, the logical translation of a set of sequences containing both $\langle +p, +q \rangle$ and $\langle +p, -q \rangle$ will contain both $q(t_p)$ and $\sim q(t_p)$, making it inconsistent in the usual syntactic sense. Analogously, to the logical translation of $\langle +p, \#q \rangle$ we should add the sentence $M[q(t_p)] \wedge M[\sim q(t_p)]$.

Theorem 2.9 Every union of distinct grounded expansions of an IS-A acyclic Γ is inconsistent.

Proof Let Φ_a and Φ_b be two distinct grounded expansions of an i.a. Γ. Let y be a minimal token such that there exists a sequence $\sigma = \langle x_1, \ldots, x_n \rangle$ in one expansion but not the other, where $y = x_{n-1}$. Choose a σ of minimal length that satisfies this

53

condition on y and assume without loss of generality that σ is in Φ_b and not in Φ_a. Both $\langle x_1, \ldots, x_{n-1} \rangle$ and $\langle x_2, \ldots, x_n \rangle$ are in Φ_a. Since σ is not, it must not be inheritable in Φ_a, which means Φ_a either contradicts or precludes it. If Φ_a contradicts it, then the union of the two expansions is inconsistent. If Φ_a precludes it, then σ has an intermediary z in Φ_a such that $\langle z, x_n' \rangle$ is in Γ. Note that $z \neq x_i$ for $1 \leq i \leq n$, since otherwise every expansion would preclude $\langle x_1, \ldots, x_n \rangle$. Thus Φ_a must contain a sequence

$$\delta = \langle x_1, \ldots, x_i, z_1, \ldots, z_m, x_{i+1} \rangle$$

(δ not in Φ_b) where z is one of $z_1 \ldots z_m$ and $1 \leq i < n$. If $i = n-1$ then $\langle z, x_n' \rangle = \langle z, x_{i+1}' \rangle$ and Φ_a would preclude δ. Therefore $i < n-1$. But then $z_m \prec x_{n-1}$, which contradicts our assumption that x_{n-1} is a minimal penultimate token for which there exists a sequence in one expansion and not the other. $\blacksquare$

I would like to extend this proof to cover the case where Γ is not i.a., but to date I have been unable to do so or to find a counterexample.

Theorem 2.10 Let Φ_a and Φ_b be grounded expansions of an IS-A acyclic Γ. Then $\Phi_a = \Phi_b$ iff $C(\Phi_a) = C(\Phi_b)$.

Proof If $\Phi_a = \Phi_b$ then $C(\Phi_a) = C(\Phi_b)$ by definition. Let $C(\Phi_a) = C(\Phi_b)$, and suppose by way of contradiction that $\Phi_a \neq \Phi_b$. Let y be a minimal token such that there exists a sequence $\sigma = \langle x_1, \ldots, x_n \rangle$ in one expansion but not the other, where $y = x_{n-1}$. Assume without loss of generality that $\sigma \in \Phi_b - \Phi_a$. Both $\langle x_1, \ldots, x_{n-1} \rangle$ and $\langle x_2, \ldots, x_n \rangle$ are in Φ_a; since σ is not, it must not be inheritable in Φ_a. Φ_a cannot contradict σ or else Φ_b would too because their conclusion sets are equal, so Φ_a must preclude σ. But by the argument used in the preceding theorem, this contradicts our assumption that y was a minimal point of discrepancy between the two expansions. $\blacksquare$

As with the previous theorem, I have to date been unable either to extend the proof to the case where Γ is not i.a. or to find a counterexample.

2.12   Existence

For the notion of grounded expansions to be useful, we need to know when they exist and how to compute them. In this section we prove that every i.a. Γ has a grounded expansion by showing how to construct one by a process of successive approximations. The question of existence in the general case remains open.

Theorem 2.11 Every IS-A acyclic Γ has a grounded expansion.

Proof Let Γ be i.a. We construct a grounded expansion of Γ by successively adding in inheritable sequences in such a way so that later additions do not contradict or preclude (and hence invalidate) prior sequences. Recall that Γ is finite since inheritance graphs are assumed to have a finite number of nodes. By Theorem 2.7, the grounded expansion we construct will also be finite.

For every set $\phi \subseteq \Sigma$, we define $I(\phi)$ to be the set of sequences that are inheritable in ϕ but not contained in ϕ. $I^*(\phi) \subseteq I(\phi)$ is defined to be the set of *minimal* inheritable sequences under the $\prec$ ordering on sequences. (The $\prec$ relation on sequences need not be a partial ordering for all ϕ, but for those ϕ's of interest to us it will be, since Γ is *i.a.* Therefore, for the ϕ's of interest to us, if $I(\phi)$ is nonempty it contains a minimal sequence.) We define $a(\phi)$, the set of augmentations of ϕ, to be the sets consisting of ϕ plus a single minimal inheritable sequence. Stated precisely, $\phi' \in a(\phi)$ iff $\phi' = \phi \cup \{\sigma\}$ for some $\sigma \in I^*(\phi)$.

We note that since the set of inheritable sequences is finite and the $\prec$ relation on sequences is a partial ordering when Γ is i.a., $a(\phi) = \emptyset$ iff $I(\phi) = \emptyset$. In other words, if any sequences are inheritable in ϕ there will be at least one minimal inheritable sequence. We also note without proof that $a(\phi)$ may easily be computed from ϕ.

A sequence of successive approximations is defined to be any sequence $\{\phi_i\}_{i=0}^{\infty}$ such that $\phi_0 = \Gamma$ and for each $i \geq 0$, $\phi_{i+1} \in a(\phi_i)$ if $a(\phi_i) \neq \emptyset$; otherwise $\phi_{i+1} = \phi_i$. We will show that $\Phi = \cup_{i=0}^{\infty} \phi_i$ is a grounded expansion of Γ, no matter what choices of augmentations and order are made in the successive approximations.

We first note that $\phi_i \subseteq \phi_{i+1} \subseteq \Phi$ for each i, thus justifying the description "successive approximation." In particular, every sequence of ϕ_i contradicted in (precluded by) ϕ_i is contradicted in (precluded by) ϕ_{i+1}. We also note that if $\phi_{i+1} = \phi_i$ then $\phi_i = \Phi$, since $\phi_{i+1} = \phi_i$ means that $a(\phi_i) = \emptyset$, and hence that $\phi_i = \phi_{i+n}$ for all $n \geq 0$.

Now Φ is clearly finite, since there are only a finite number of inheritable sequences when Γ is *i.a.* Therefore only a finite number of successive approximations can be augmentations. This means that for some $i \geq 0$, $\phi_i = \phi_{i+1} = \Phi$, so $|\Phi| \leq |\Gamma| + i$. (In fact, $|\Phi| = |\Gamma| + j$, where j is the least i such that $\phi_i = \phi_{i+1}$.)

Φ is also clearly closed under inheritance, since $\Phi = \phi_i$ means that $\phi_i = \phi_{i+1}$, hence $a(\phi_i) = \emptyset$, hence $I(\phi_i) = \emptyset$, hence $I(\Phi) = \emptyset$.

Now we must show that successive augmentations do not invalidate prior ones. There are several steps to this. First, we show that if $\phi_{i+1} = \phi_i \cup \{\sigma\}$ then for all $\tau \in I^*(\phi_j)$, $j > i$, $\tau \not\prec \sigma$. The proof is by contradiction. Suppose τ is a sequence of minimal length such that that $\tau \prec \sigma$ and $\tau \in I^*(\phi_j)$ for some $j > i$. Note that since $\tau \prec \sigma$, $\tau \notin I^*(\phi_i)$ because σ is by definition a minimal sequence inheritable in ϕ_i; σ would not be minimal if τ were inheritable in ϕ_i. But $\tau \in I^*(\phi_j)$, so τ cannot be contradicted or precluded by ϕ_i. Therefore ϕ_i contains at most one of the two subsequences of τ, call them τ_1 and τ_2, that must be present in ϕ_j for τ to be in $I^*(\phi_j)$. Let $\tau_1 = \langle x_1, \ldots, x_{n-1} \rangle$ and $\tau_2 = \langle x_2, \ldots, x_n \rangle$. For some k, $i \leq k < j$, $\phi_{k+1} = \phi_k \cup \{\tau_1\}$ or $\phi_{k+1} = \phi_k \cup \{\tau_2\}$. Note that $\tau_1 \prec \sigma$ and $\tau_2 \prec \sigma$, and both τ_1 and τ_2 are shorter than τ. This contradicts our assumption that τ was a sequence of minimal length falsifying the assertion.

Second, we show that each sequence of ϕ_i contradicted in ϕ_{i+1} is contradicted in ϕ_i. If ϕ_{i+1} contradicts a sequence $\sigma \in \phi_i$ that is not contradicted in ϕ_i, then $\phi_{i+1} = \phi_i \cup \{\tau\}$ for some $\tau \in I^*(\phi_i)$ such that τ contradicts σ. But then ϕ_i would contradict τ, so τ could not be inheritable in ϕ_i.

Third, we show that each sequence of ϕ_i precluded in ϕ_{i+1} is precluded in ϕ_i. If ϕ_{i+1} precludes a sequence $\sigma = \langle x_1, \ldots, x_n \rangle \in \phi_i$ that is not precluded in ϕ_i, then there exists some y such that Γ contains $\langle y, x_n' \rangle$ and y is an intermediary to σ in ϕ_{i+1} but not in ϕ_i. Since y must become an intermediary to σ in ϕ_{i+1}, we see that

$$\phi_{i+1} = \phi_i \cup \{\delta\}$$

where $\delta = \langle x_1, \ldots, x_j, z_1, \ldots, z_m, x_{j+1} \rangle$, y is one of $z_1 \cdots z_m$, and $1 \leq j < n - 1$. But $\delta \prec \sigma$, so $\delta \notin I^*(\phi_i)$.

To complete the proof that successive augmentations do not invalidate prior ones, we show that every $\sigma \in \phi_i - \phi_0$ is inheritable in ϕ_{i+1}. Note that if σ is inheritable in ϕ_i then it is not contradicted or precluded in ϕ_i. By the previous steps, if σ is contained in ϕ_i and not contradicted or precluded in ϕ_i then it is not contradicted or precluded in ϕ_{i+1}. Also note that if $\sigma = \langle x_1, \ldots, x_n \rangle$ is inheritable in ϕ_i then ϕ_i contains $\langle x_1, \ldots, x_{n-1} \rangle$ and $\langle x_2, \ldots, x_n \rangle$, which means ϕ_{i+1} does too. Thus, σ is inheritable in ϕ_{i+1}. By transitivity, if $\sigma \in \phi_i$ is inheritable in ϕ_i then it is inheritable in Φ.

Finally, to show that Φ is a grounded expansion of Γ, we first note that $\phi_0 = \Gamma$. Since $\Phi \supseteq \Gamma$ and Φ is closed under inheritance, Φ is an expansion of Γ. Then, since every element of $\Phi - \Gamma$ is inheritable in Φ, we see that Φ is a grounded expansion of Γ.

∎

Since the proof is by construction, we have the following corollary:

Corollary 2.8 Every IS-A acyclic Γ has a constructable grounded expansion.

The requirement that Γ be i.a. in the above proof may be related to Etherington's results on the existence of constructable expansions (extensions in his terminology) for hierarchical ordered seminormal default theories (Etherington, 1983). Etherington proved that every ordered seminormal default theory has at least one expansion. He also gave a procedure for constructing one of these expansions when the theory is hierarchical. But his ordering is a dependency ordering: every normal theory is ordered in his terminology, and his use of the term "hierarchical" does not imply acyclicity. Therefore Etherington's result might be stronger than mine, but the two are not directly comparable since he does not use an inferential distance ordering to effect exceptions. Instead, as noted previously, he handles exceptions explicitly by making the defaults seminormal.

2.13 Ambiguity

Some graphs have a single grounded expansion, while others have several. For example, we have seen that the graphs in figure 2.6 each have one expansion while the one in figure 2.7 has two. Compare the two graphs in figure 2.8. In 2.8a, Γ contains $\langle +b, -d \rangle$ and $+b$ is an intermediary to $\langle +a, +c, +d \rangle$, so $\langle +a, +c, +d \rangle$ is precluded. This graph has a single expansion. In 2.8b, $+b$ is not an intermediary to any sequence containing $+c$, so both $\langle +a, +c, +d \rangle$ and $\langle +a, +b, -d \rangle$ are inheritable. These two sequences contradict each other. That is why figure 2.8b has two expansions. This graph can be redrawn as figure 2.8c, which makes the reason for the multiple expansions more apparent.

Definition 2.16 A set of ordered pairs Γ is *ambiguous* iff it has more than one grounded expansion. Γ is *unambiguous* iff it is not ambiguous.

Theorem 2.12 The expansion of every unambiguous IS-A acyclic Γ is constructable.

Proof Every i.a. Γ has a constructable grounded expansion. If Γ is unambiguous then it has only one expansion, so the expansion we can construct is the only one there is.

∎

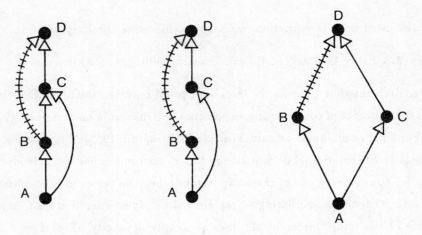

Figure 2.8: (a) a network with a single grounded expansion; (b) a network with two grounded expansions; (c) another view of (b).

Ambiguity is a different kind of property than consistency. The consistency of a grounded expansion Φ of Γ is a strictly local property of Γ, because to show that Φ is consistent or inconsistent we need only look at pairs of sequences in Γ. If we find a pair of form $\langle x, y \rangle$, $\langle x, y' \rangle$ or a single sequence of form $\langle x, x' \rangle$, then Γ and Φ are inconsistent. Ambiguity is a more global property: it is determined not by the existence of particular sequences in Γ or in any one expansion, but by the existence of multiple expansions.

In this section we will define a property of individual expansions called *stability* and show that a grounded expansion of a totally acyclic Γ is stable iff Γ is unambiguous. Stability is a more local property than ambiguity, since we can tell whether an acyclic Γ is ambiguous by examining just one of its expansions for stability. But stability is still more global than consistency, because to show that an expansion is unstable we must show that certain sequences are *not* precluded in it. Thus there is no minimum number of sequences we can examine to determine stability; stability depends on the entire expansion.

Definition 2.17 An expansion Φ is *unstable* iff it contains a set of sequences of form $\langle x, y_1, \ldots, y_n \rangle$, $\langle x, z_1, \ldots, z_m \rangle$, $\langle y_1, \ldots, y_n, w \rangle$, and $\langle z_1, \ldots, z_m, w' \rangle$, such that Φ precludes neither $\langle x, y_1, \ldots, y_n, w \rangle$ nor $\langle x, z_1, \ldots, z_m, w' \rangle$. Φ is *stable* iff it is not unstable. See figure 2.9.

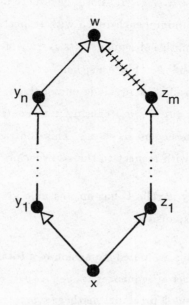

Figure 2.9: The subgraph that causes instability.

Theorem 2.13 If a totally acyclic Γ is ambiguous then each of its grounded expansions is unstable.

Proof Let Φ_a and Φ_b be two distinct grounded expansions of an ambiguous totally acyclic Γ. We will show that Φ_b is unstable. Let w be a minimal token with respect to $\ll$, and let σ be a sequence of minimal length given w, such that $\sigma = \langle x, y_1, \ldots, y_n, w \rangle$ appears in Φ_a but not Φ_b. Then both $\langle x, y_1, \ldots, y_n \rangle$ and $\langle y_1, \ldots, y_n, w \rangle$ are in Φ_b. Since σ is not in Φ_b, σ must not be inheritable in Φ_b, which means Φ_b either contradicts or precludes σ. Suppose Φ_b contradicts σ. Since Φ_b contains $\langle x, y_1, \ldots, y_n \rangle$ and is grounded in Γ, it cannot contradict $\langle x, y_1, \ldots, y_n \rangle$, so the only way Φ_b can contradict σ is if $C(\Phi_b)$ contains $\langle x, w' \rangle$, which means Φ_b contains some sequence $\langle x, z_1, \ldots, z_m, w' \rangle$. Therefore Φ_b also contains $\langle x, z_1, \ldots, z_m \rangle$ and $\langle z_1, \ldots, z_m, w' \rangle$, making Φ_b unstable.

Now suppose that Φ_b precludes σ instead of contradicting it. Then there exists an intermediary v to σ such that $\langle v, w' \rangle$ is in Γ. Note that $v \ll w$. The fact that v is an intermediary to σ in Φ_b but not Φ_a implies there exists a sequence $\langle t_1, \ldots, t_p \rangle$ in Φ_b but not in Φ_a, where $t_1 = x$, $t_i = v$ for some i, and $t_p \ll w$; this sequence establishes v as an intermediary to σ. We will show that because this sequence is not in Φ_a, Φ_a must contain some other sequence $\langle u_1, \ldots, u_k \rangle$ not in Φ_b where $u_k \ll w$, which contradicts our assumption that w was a minimal such token with respect to the $\ll$ ordering.

Consider a maximal length subsequence $\langle t_i, \ldots, t_j \rangle$ of $\langle t_1, \ldots, t_p \rangle$, $1 \leq i < j \leq p$, that appears in both Φ_a and Φ_b. Since neither $\langle t_{i-1}, \ldots, t_j \rangle$ nor $\langle t_i, \ldots, t_{j+1} \rangle$ appears in Φ_a, Φ_a must either contradict or preclude one of them, while Φ_b does not. But then Φ_a contains some sequence $\langle u_1, \ldots, u_k \rangle$ causing it to contradict or preclude $\langle t_i, \ldots, t_j \rangle$, where $\langle u_1, \ldots, u_k \rangle$ is not in Φ_b and $u_k \ll w$. This contradicts our assumption that w was a minimal such token with respect to the $\ll$ ordering. ∎

Theorem 2.14 If a totally acyclic Γ has an unstable grounded expansion then it has more than one grounded expansion.

Proof Let Φ be an unstable grounded expansion of a totally acyclic Γ, that is, let Φ be an expansion containing a set of sequences $\langle x, y_1, \ldots, y_n \rangle$, $\langle x, z_1, \ldots, z_m \rangle$, $\langle y_1, \ldots, y_n, w \rangle$, and $\langle z_1, \ldots, z_m, w' \rangle$, such that Φ precludes neither $\langle x, y_1, \ldots, y_n, w \rangle$ nor $\langle x, z_1, \ldots, z_m, w' \rangle$. We define the *free choice set* for Φ with respect to x and w, written $\Delta_x^w(\Phi)$, to be the sequences of Φ except those of length greater than two that contain x or any token $\ll x$, and also contain w, w', or any token $\gg w$. Note that $\Gamma \subseteq \Delta_x^w(\Phi)$. Let Ψ be a grounded expansion of $S = \Delta_x^w(\Phi) \cup \{\langle x, y_1, \ldots, y_n, w \rangle\}$. Then we can make the following observations.

Every sequence in $\Psi - \Delta_x^w(\Phi)$ contains x or a token $\ll x$, and also contains w, w', or a token $\gg w$. Proof: Suppose by way of contradiction that $\langle v_1, \ldots, v_p \rangle$ is a sequence of minimal length in $\Psi - \Delta_x^w(\Phi)$ that contains neither x nor a token $\ll x$, or else contains neither w, w', nor a token $\gg w$. Ψ must contain $\langle v_1, \ldots, v_{p-1} \rangle$ and $\langle v_2, \ldots, v_p \rangle$ as well. If both these sequences were in $\Delta_x^w(\Phi)$ then either $\langle v_1, \ldots, v_p \rangle$ or some contradictory sequence is in Φ, and hence in $\Delta_x^w(\Phi)$, in which case $\langle v_1, \ldots, v_p \rangle$ could not be in $\Psi - \Delta_x^w(\Phi)$. Therefore at least one of $\langle v_1, \ldots, v_{p-1} \rangle$ or $\langle v_2, \ldots, v_p \rangle$ must be in $\Psi - \Delta_x^w(\Phi)$. This contradicts our assumption that $\langle v_1, \ldots, v_p \rangle$ was chosen to be of minimal length.

Ψ cannot preclude $\langle x, y_1, \ldots, y_n, w \rangle$. Proof: If Ψ precludes $\langle x, y_1, \ldots, y_n, w \rangle$ then so does $\Delta_x^w(\Phi)$, because no sequence in $\Psi - \Delta_x^w(\Phi)$ can establish the preclusion. But if $\Delta_x^w(\Phi)$ precludes the sequence then so must Φ, and we know that Φ does not. Therefore Ψ cannot.

60

Ψ *is a grounded expansion of* $\Delta_x^w(\Phi)$. Proof: Ψ is a grounded expansion of $S = \Delta_x^w(\Phi) \cup \{\langle x, y_1, \ldots, y_n, w \rangle\}$. To ground Ψ in $\Delta_x^w(\Phi)$ we must show that $\langle x, y_1, \ldots, y_n, w \rangle$ is inheritable in Ψ. $\Delta_x^w(\Phi)$ contains $\langle x, y_1, \ldots, y_n \rangle$ and $\langle y_1, \ldots, y_n, w \rangle$, and does not preclude $\langle x, y_1, \ldots, y_n, w \rangle$. $\Delta_x^w(\Phi)$ also does not contradict $\langle x, y_1, \ldots, y_n, w \rangle$, because any contradictory sequence has been removed from it. Therefore $\langle x, y_1, \ldots, y_n, w \rangle$ is inheritable in Ψ, so Ψ is grounded in $\Delta_x^w(\Phi)$.

Ψ *is a grounded expansion of* Γ. Proof: Ψ is an expansion of $S = \Delta_x^w(\Phi) \cup \{\langle x, y_1, \ldots, y_n, w \rangle\}$, which is a superset of Γ, so Ψ is an expansion of Γ. To *ground* Ψ in Γ we must show that every element of $\Psi - \Gamma$ is inheritable in Ψ. Clearly the elements of $\Psi - S$ are inheritable in Ψ; it remains to show that the elements of $S - \Gamma$ are inheritable in Ψ. By the previous lemma, $\langle x, y_1, \ldots, y_n, w \rangle$ is inheritable in Ψ, so the problem is further reduced to showing that the elements of $\Delta_x^w(\Phi) - \Gamma$ are inheritable in Ψ.

Suppose by way of contradiction that there exists a sequence $\langle v_1, \ldots, v_p \rangle$ in $\Delta_x^w(\Phi) - \Gamma$ that becomes uninheritable in Ψ. Then either $v_1 = x$ or $v_1 \ll x$, since $\Delta_x^w(\Phi)$ is closed under inheritance with respect to other types of sequences. Consequently, $v_p \ll w$. Since Ψ precludes $\langle v_1, \ldots, v_p \rangle$ while $\Delta_x^w(\Phi)$ does not, Γ must contain a sequence $\langle u, v_p' \rangle$ such that u is an intermediary to $\langle v_1, \ldots, v_p \rangle$ in Ψ but not in $\Delta_x^w(\Phi)$. This tells us that $u \neq v_i$ for $1 \leq i \leq p$.

Let $\langle t_1, \ldots, t_q \rangle$ be a sequence of $\Psi - S$ making u an intermediary to $\langle v_1, \ldots, v_p \rangle$ in Ψ. Then $t_1 = v_1$ and $t_q = v_i$ for some $i, 1 \leq i < p - 1$, so $t_q \ll v_p$ and hence $t_q \ll w$. But then $\langle t_1, \ldots, t_q \rangle$ cannot be in $\Psi - S$.

Γ *has at least two grounded expansions.* Proof: Φ is an unstable grounded expansion of Γ containing sequences $\langle x, y_1, \ldots, y_n \rangle$, $\langle x, z_1, \ldots, z_m \rangle$, $\langle y_1, \ldots, y_n, w \rangle$, and $\langle z_1, \ldots, z_m, w' \rangle$, such that Φ precludes neither $\langle x, y_1, \ldots, y_n, w \rangle$ nor $\langle x, z_1, \ldots, z_m, w' \rangle$. Construct a grounded expansion Ψ_a of $\Delta_x^w(\Phi) \cup \{\langle x, y_1, \ldots, y_n, w \rangle\}$, and a grounded expansion Ψ_b of $\Delta_x^w(\Phi) \cup \{\langle x, z_1, \ldots, z_m, w' \rangle\}$. Ψ_a and Ψ_b are both grounded expansions of Γ. Since one contains an inheritable sequence contradicted by the other, the two expansions must be distinct. Hence, Γ is ambiguous, and the proof is complete. ∎

Corollary 2.9 An acyclic Γ is unambiguous iff it has a stable grounded expansion.

We have seen that if an acyclic Γ has multiple grounded expansions then each is unstable. Conversely, if an acyclic Γ has one unstable grounded expansion then it has several expansions. Thus, at least for acyclic graphs, we can tell whether Γ is ambiguous by constructing one expansion and checking it for stability.

Etherington has pointed out in personal communication that the requirement that Γ be totally acyclic can be relaxed somewhat. For example, a network containing $\langle +a, -b \rangle$, $\langle +b, -c \rangle$, and $\langle +c, -a \rangle$ is cyclic (though it may still be i.a.), but this cycle cannot generate any cyclic inheritance paths. Thus, the strengthening of the results in this section is a possibility.

2.14   Independence of consistency and ambiguity

To show that consistency and ambiguity are orthogonal, we exhibit trivial Γ's that are (i) consistent and unambiguous; (ii) inconsistent and unambiguous; (iii) consistent and ambiguous; and (iv) inconsistent and ambiguous. See figure 2.10.

$$\text{Consistent, Unambiguous:} \quad \Gamma_1 = \Phi_1 = \{\langle +a, +b \rangle\}$$

$$\text{Inconsistent, Unambiguous:} \quad \Gamma_2 = \Phi_2 = \{\langle +a, +b \rangle, \langle +a, -b \rangle\}$$

$$\text{Consistent, Ambiguous:} \quad \Gamma_3 = \{\langle +c, +d \rangle, \langle +d, +f \rangle, \langle +c, +e \rangle, \langle +e, -f \rangle\}$$
$$\Phi_{3a} = \Gamma_3 \cup \{\langle +c, +d, +f \rangle\}$$
$$\Phi_{3b} = \Gamma_3 \cup \{\langle +c, +e, -f \rangle\}$$

$$\text{Inconsistent, Ambiguous:} \quad \Gamma_4 = \Gamma_3 \cup \Gamma_2$$
$$\Phi_{4a} = \Phi_{3a} \cup \Phi_2$$
$$\Phi_{4b} = \Phi_{3b} \cup \Phi_2$$

2.15   Size

Let N denote the number of nodes in a well-formed inheritance graph, and let L denote the number of links, i.e. $N = |\Pi|$ and $L = |\Gamma|$. Then $0 \leq L \leq 3N^2$, since we can have up to N^2 links of each of three types: IS-A, IS-NOT-A, and NO-CONCLUSION. For i.a. networks, we can also bound the size and the number of expansions.

Theorem 2.15 The size of a grounded expansion of an IS-A acyclic Γ is $O(2^N)$.

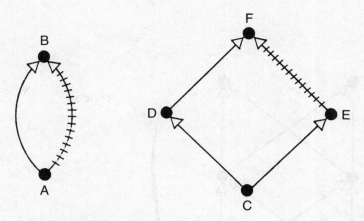

Figure 2.10: Network illustrating the independence of consistency and ambiguity.

Proof   There are N nodes in the graph, each with a corresponding positive token. Each sequence in the expansion either contains a particular positive token or it does not. Since Γ is i.a., no token appears twice in any sequence. Thus there are at most 2^N sequences consisting of positive tokens. The expansion may contain other sequences, but only the last token in a sequence can be negative or neutral, so the total number of sequences is no more than three times the number of strictly positive sequences.   ∎

We can show that this bound cannot be reduced further. Consider a graph of N nodes a_1 through a_N. Let $\Gamma = \{\langle +a_i, +a_j \rangle \mid 1 \le i < j \le N\}$. The grounded expansion of this i.a. graph contains $2^N - N - 1$ sequences, which is $O(2^N)$. But to construct the graph requires $O(N^2)$ links. Figure 2.11 shows an i.a. construction due to Jim Saxe that achieves an expansion size of $2^{N/2}$ sequences using only $4(N-1)$ links. For the case where Γ is not i.a., meaning the IS-A hierarchy contains cycles, there is no bound on the size of a grounded expansion.

Theorem 2.16 The number of distinct grounded expansions of an IS-A acyclic Γ is at most 3^{N^2}.

Proof   There are at most 3^{N^2} distinct conclusion sets $C(\Phi)$. When Γ is i.a., $\Phi_a = \Phi_b$ if $C(\Phi_a) = C(\Phi_b)$, so there are at most that many distinct grounded expansions.   ∎

Figure 2.12 shows an i.a. construction that achieves $3^{N^2/4}$ distinct grounded expansions. Although the size of an expansion and the number of sequences it contains can

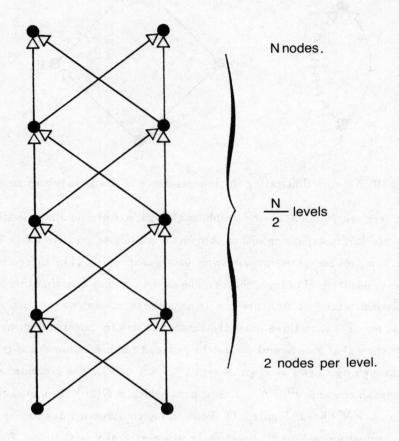

N nodes.

$\dfrac{N}{2}$ levels

2 nodes per level.

Figure 2.11: A network whose grounded expansion contains $2^{N/2}$ sequences.

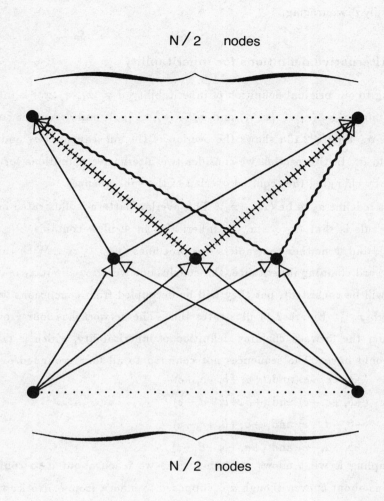

N / 2   nodes

N / 2   nodes

Figure 2.12: A network with $3^{N^2/4}$ grounded expansions.

65

be very large in the worst case, as the pathological networks in figures 2.11 and 2.12 demonstrate, in actual knowledge representation applications the sizes need not be computationally discouraging.

2.16 Alternative definitions for inheritability

According to our original definition of inheritability, $\sigma = \langle x_1, \ldots, x_n \rangle$ is inheritable in Φ iff Φ contains $\sigma_1 = \langle x_1, \ldots, x_{n-1} \rangle$ and $\sigma_2 = \langle x_2, \ldots, x_n \rangle$, and Φ neither contradicts nor precludes σ. Figure 2.13a shows the overlap of the subsequences σ_1 and σ_2 and their relation to σ. In this section we consider two alternative definitions for inheritability obtained by changing the required overlap of the subsequences.

Let us redefine σ_2 to be $\langle x_{n-1}, x_n \rangle$. This overlap pattern is illustrated in figure 2.13b. Our new rule is that $\langle x_1, \ldots, x_n \rangle$ is inheritable in Φ iff Φ contains $\langle x_1, \ldots, x_{n-1} \rangle$ and $\langle x_{n-1}, x_n \rangle$, and Φ neither contradicts nor precludes $\langle x_1, \ldots, x_n \rangle$. With this rule, which I call forward chaining inheritance, the conclusions we draw about x_1 in any single expansion will be consistent, but they will be decoupled from conclusions we draw about x_2 through x_{n-1}. Figure 2.14 illustrates this. The network has four grounded expansions under the forward chaining definition of inheritability, which is twice the number it should have. The sequences not common to all four grounded expansions are:

Φ_1: $\langle +b, +c, +e \rangle$ and $\langle +a, +b, +c, +e \rangle$

Φ_2: $\langle +b, +c, +e \rangle$ and $\langle +a, +b, +d, -e \rangle$

Φ_3: $\langle +b, +d, -e \rangle$ and $\langle +a, +b, +c, +e \rangle$

Φ_4: $\langle +b, +d, -e \rangle$ and $\langle +a, +b, +d, -e \rangle$

Decoupling is what allows the conclusions we reach about a to conflict with our conclusions about b, even though a is supposed to inherit from b. Notice that in Φ_3 the sequence $\langle +a, +b, +c, +e \rangle$ appears in an expansion containing $\langle +b, +d, -e \rangle$. According to the forward chaining rule, all that is required for $\langle +a, +b, +c, +e \rangle$ to be inheritable is that the expansion contain $\langle +a, +b, +c \rangle$ and $\langle +c, +e \rangle$, and neither contradict nor preclude $\langle +a, +b, +c, +e \rangle$.

With double chaining inheritance, on the other hand, $\langle +a, +b, +c, +e \rangle$ would only be inheritable in an expansion containing $\langle +a, +b, +c \rangle$ and $\langle +b, +c, +e \rangle$. Similarly, $\langle +a, +b, +d, -e \rangle$ would only be inheritable in an expansion containing $\langle +a, +b, +d \rangle$ and $\langle +b, +d, -e \rangle$. Thus under the double chaining rule, Φ_1 and Φ_4 would be legal grounded

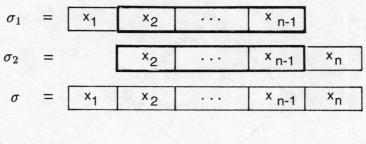

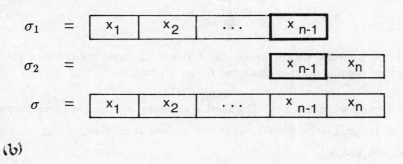

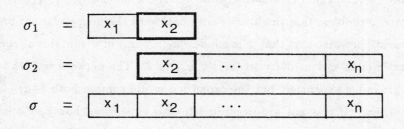

Figure 2.13: Inheritability defined by (a) double chaining as used in this thesis; (b) forward chaining; (c) backward chaining.

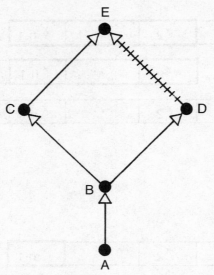

Figure 2.14: A network which, under the forward chaining inheritance rule, would have four grounded expansions due to decoupling.

expansions but Φ_2 and Φ_3 would not. The double chaining rule makes sense because if a is inheriting all its properties from b, whatever we conclude about a ought to agree with our conclusions about b.

Figure 2.13c illustrates another alternative to double chaining inheritance which I call backward chaining. Here the rule is that $\langle x_1, \ldots, x_n \rangle$ is inheritable in Φ iff Φ contains $\langle x_1, x_2 \rangle$ and $\langle x_2, \ldots, x_n \rangle$, and Φ neither contradicts nor precludes $\langle x_1, \ldots, x_n \rangle$. Figure 2.15 illustrates a problem that would arise with backward chaining under our present definition of preclusion which says that Φ precludes $\langle x_1, \ldots, x_n \rangle$ iff Φ contains a sequence $\langle z, x_n' \rangle$ such that z is an intermediary to $\langle x_1, \ldots, x_n \rangle$ in Φ. The network in 2.15 would have a unique grounded expansion, but the expansion would contain both $\langle +a, +b, -d \rangle$ and $\langle +a, +c, +d, +e \rangle$. If a is not a d according to this expansion, how can a inherit properties via d? The problem arises because, under the present definition of preclusion, Φ precludes $\langle +a, +c, +d \rangle$ but not $\langle +a, +c, +d, +e \rangle$. Under the backward chaining rule, $\langle +a, +c, +d, +e \rangle$ is inheritable because Φ contains $\langle +a, +c \rangle$ and $\langle +c, +d, +e \rangle$; it does not matter than Φ does not contain $\langle +a, +c, +d \rangle$. We could fix things by changing the definition of preclusion as follows: Φ shall preclude $\langle x_1, \ldots, x_n \rangle$ iff for some x_i, $1 \leq i \leq n$, Φ contains a sequence $\langle z, x_i' \rangle$ such that z is an intermediary to $\langle x_1, \ldots, x_i \rangle$ in Φ.

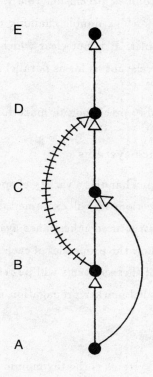

Figure 2.15: A network illustrating why the present definition of preclusion would have to be revised if backward chaining inheritance were used.

Other than the fact that it would complicate the definition of preclusion slightly, I have found no reason to object to the backward chaining version of the inheritance rule, and offer two conjectures. First, I conjecture that under the modified preclusion rule, backward chaining and double chaining are equivalent. Second, I conjecture that backward chaining with the modified preclusion rule is equivalent to double chaining with the original preclusion rule. The double chaining definition of inheritance makes some of the proofs slightly simpler. It is not clear which definition gives an easier Lisp mechanization. The choice is irrelevant as far as parallel marker propagation algorithms are concerned.

I have found no way to modify preclusion to make forward chaining work correctly.

2.17 Specialized inheritance systems

The general theory have developed handles a variety of specialized inheritance systems as special cases. In the following sections we will examine taxonomic hierarchies, exception-free inheritance systems, tree-structured inheritance systems, and orthogonal multiple inheritance systems, and show how the properties of each can be derived from the general theory. The characterizations of these systems will be referred to in chapter 4 where we consider the correctness of parallel marker propagation machine algorithms.

2.18 Taxonomic hierarchies

The simplest of all inheritance systems is the taxonomic hierarchy, where each class has at most one immediate superclass. This type of inheritance system has been common since the earliest AI programs; taxonomic classification itself is a method of organizing knowledge that predates the development of computers.

We can formally define our notion of a taxonomic hierarchy by giving the conditions under which an inheritance graph qualifies as one:

Definition 2.18 Γ is a taxonomic hierarchy iff Γ mentions only positive tokens, and $\langle x, y \rangle, \langle x, z \rangle \in \Gamma$ implies $y = z$.

In graph terminology, a taxonomic hierarchy is a forest of rooted trees containing only nodes and IS-A links. Since taxonomic hierarchies are a special case of class/property inheritance systems, their properties are further discussed in the following section.

2.19 Class/property inheritance systems

A class/property inheritance system consists of a set of classes with a strictly transitive inheritance relation and a set of inheritable properties associated with each class. See figure 2.16. The inheritance of properties, but not the inheritance of class membership, is subject to exceptions. Frame-based systems such as FRL are class/property systems. In FRL, the frames make up the classes, which are connected by inviolable AKO links. Properties are expressed as slot values, and are subject to exceptions as described in chapter 1.

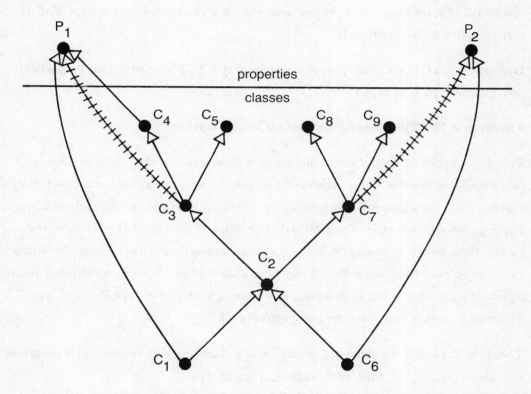

Figure 2.16: An orthogonal class/property inheritance system. Nodes C_1 through C_9 are classes; P_1 and P_2 are properties.

In a *tree-structured* class/property inheritance system, each class may inherit from at most one immediate superclass. Classes in Simula and Smalltalk-76, and derived types in Ada are implemented as inheritance systems of this type. To formally define the tree-structured class/property inheritance system in terms of our general theory of

71

inheritance, we first define two types of predicate nodes. Nodes of type C are those that can be viewed as representing classes, while nodes of type P are those that can be viewed as representing properties.

Definition 2.19 A predicate node $y \in \Pi$ is of type C (denotes a class) with respect to Γ iff no sequence in Γ ends with $-y$ or $\#y$. A predicate *token* is of type C iff the corresponding node is of type C.

Definition 2.20 A predicate node $y \in \Pi$ is of type P (denotes a property) with respect to Γ iff no sequence in Γ begins with $+y$. A predicate *token* is of type P iff the corresponding node is of type P.

Definition 2.21 Γ is a class/property system iff it is IS-A acyclic and all its predicate nodes are of type P or type C.

Theorem 2.17 Class/property systems are totally acyclic.

Proof Suppose by way of contradiction that there exists a class/property system Γ containing two nodes x and y such that $x \ll y$ and $y \ll x$, meaning Γ was not totally acyclic. Since class/property systems are by definition IS-A acyclic, the cycle involving x and y must contain at least one IS-NOT-A or NO-CONCLUSION link. Let u, v, and w be any three nodes on this cycle (not necessarily distinct) such that Γ contains either $\langle +u, -v \rangle$ or $\langle +u, \#v \rangle$, and a link of any type from v to w. Node v, which must be a predicate node since it has an incoming link from u, is neither of type P nor of type C. Therefore Γ cannot be a class/property system. ∎

Theorem 2.18 If Φ is a grounded expansion of a class/property system and Φ contains a sequence $\langle x_1, \ldots, x_n \rangle$, then x_2 through x_{n-1} are of type C.

Proof By induction on n. ∎

Definition 2.22 Γ is *a tree-structured class/property system* iff it is a class/property system and, for all nodes x, y and z of type C, if $\langle +x, +y \rangle \in \Gamma$ and $\langle +x, +z \rangle \in \Gamma$ then $y = z$.

Corollary 2.10 Taxonomic hierarchies are tree-structured class/property systems.

The network in figure 2.17a is tree-structured under the above definition, but the one in figure 2.17b is not because some nodes inherit from more than one superior. Figure 2.18 shows one branch of a tree-structured class/property system with five classes, C_1 through C_5, forming a chain, and two properties, P_1 and P_2, about which the classes disagree and make exceptions.

Tree structured class/property systems are always unambiguous and have unique, constructable grounded expansions. Rather than prove this result directly, I will prove a stronger one by considering the case of multiple inheritance.

We can easily generalize the concept of a class/property system to allow for multiple inheritance. However, if this is done in an unrestricted way, the resulting system will be vulnerable to ambiguity and confusion caused by redundancy in the same way as the fully general inheritance system analyzed earlier. Instead, we will allow multiple inheritance in a restricted form by introducing the notion of an *orthogonal* class/property system. The idea is to allow a class to inherit from multiple superclasses provided that the properties it inherits from any two superclasses are disjoint.

Definition 2.23 Γ is an *orthogonal* class/property system iff it is a class/property system and, for every quartet of distinct nodes x, y, z, and w, if $\langle +x, +y \rangle \in \Gamma$ and $\langle +x, +z \rangle \in \Gamma$ then either $y \not\ll w$ or $z \not\ll w$.

Corollary 2.11 If Γ is tree-structured then it is orthogonal.

Theorem 2.19 If Γ is orthogonal then it has a constructable grounded expansion.

Proof Earlier we proved that every i.a. Γ has a constructable grounded expansion. If Γ is orthogonal then it is i.a. ∎

Lemma 2.1 Let $\langle x, y_1, \ldots, y_n, w \rangle$ and $\langle x, z_1, \ldots, z_m, w' \rangle$ be sequences in a grounded expansion of an orthogonal class/property system, where $n \leq m$. Then $y_i = z_i$ for $1 \leq i \leq n$.

Proof By induction on the lengths of the sequences. ∎

Theorem 2.20 If Γ is an orthogonal class/property system then it is unambiguous.

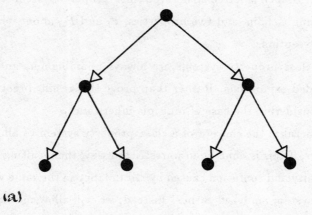

(a)

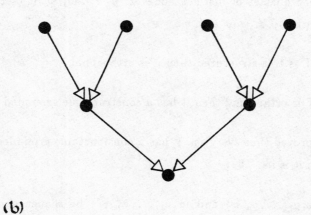

(b)

Figure 2.17: (a) a tree-structured inheritance system; (b) a system that is not tree-structured, due to the use of multiple inheritance.

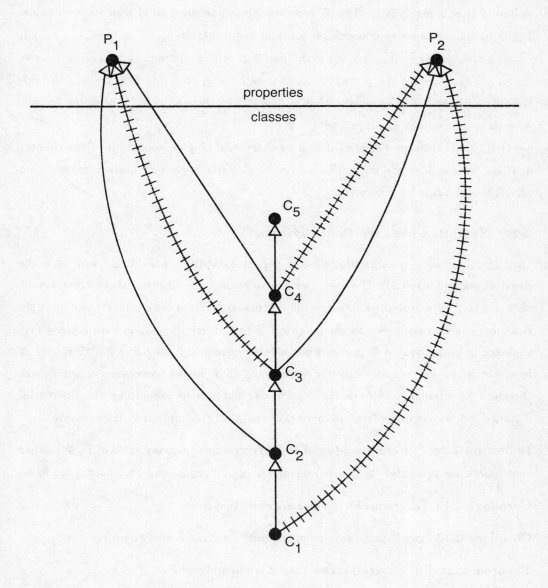

Figure 2.18: One branch of a tree-structured orthogonal class/property inheritance system.

Proof Suppose by way of contradiction that there exists an orthogonal class/property system Γ that is ambiguous. Then Γ has multiple expansions, all of which are unstable. Let Φ be one of those expansions. Φ contains sequences $\langle x, y_1, \ldots, y_n \rangle$, $\langle y_1, \ldots, y_n, w \rangle$, $\langle x, z_1, \ldots, z_m \rangle$, and $\langle z_1, \ldots, z_m, w' \rangle$ such that Φ precludes neither $\langle x, y_1, \ldots, y_n, w \rangle$ nor $\langle x, z_1, \ldots, z_m, w' \rangle$. By the preceding lemma we have $y_i = z_i$ for $1 \leq i \leq min(n, m)$. Assume without loss of generality that $n \leq m$. Then since $y_n = z_n$, Γ contains $\langle z_n, w \rangle$, so Φ does preclude $\langle x, z_1, \ldots, z_m, w' \rangle$. ∎

Orthogonal class/property inheritance systems will play an important role in chapter 4 where we consider the conditions under which certain parallel marker propagation algorithms can function correctly.

2.20   Exception-free inheritance systems

Usually inheritance systems that do not permit exceptions provide no way to make negative statements at all. However, there is no reason why negative statements cannot be permitted in a system that is free of exceptions, just as inheritance from multiple superiors can be permitted. In the language of the generic theory, any sequence whose presence in an expansion Φ prevents some other sequence from being inheritable in Φ is acting as an exception. Negative statements that do not generate exceptions are therefore permissible. We will see that in the absence of exceptions the inferential distance ordering is unnecessary; inheritance collapses to simple transitive closure.

Definition 2.24 Γ is *exception-free* iff in every grounded expansion Φ of Γ, Φ neither contradicts nor precludes $\langle x_1, \ldots, x_n \rangle$ whenever $\langle x_1, \ldots, x_{n-1} \rangle$ and $\langle x_2, \ldots, x_n \rangle$ are in Φ.

Corollary 2.12 Taxonomic hierarchies are exception-free.

Corollary 2.13 Any Γ that contains only positive tokens is exception-free.

Theorem 2.21 If Γ is exception-free then it is unambiguous.

Proof Let Φ_a and Φ_b be grounded expansions of an exception-free Γ. Let $\sigma = \langle x_1, \ldots, x_n \rangle$ be an element of minimal length in $\Phi_b - \Phi_a$. Then $\langle x_1, \ldots, x_{n-1} \rangle$ and $\langle x_2, \ldots, x_n \rangle$ are in Φ_b and, since both are shorter than σ, they are both in Φ_a as well. But since Γ is exception-free, $\langle x_1, \ldots, x_n \rangle$ must appear in Φ_a. Therefore, $\Phi_b - \Phi_a$ is empty, so $\Phi_a = \Phi_b$. ∎

Theorem 2.22 If Γ is exception-free then its grounded expansion is constructable.

Proof   We present a procedure for constructing an expansion of any exception-free Γ. Unlike in the earlier constructability proof, here we do not require Γ to be *i.a.* Let $\phi_0 = \Gamma$. Let $\langle x_1, \ldots, x_{n-1} \rangle$ and $\langle x_2, \ldots, x_n \rangle$ be sequences in ϕ_i such that $\sigma = \langle x_1, \ldots, x_n \rangle$ is not in ϕ_i. Let ϕ_{i+1} be $\phi_i \cup \{\sigma\}$ if such a σ exists, else simply ϕ_i. Let $\Phi = \cup_{i=0}^{\infty} \phi_i$. Then Φ is a superset of Γ and closed under inheritance, so Φ is an expansion of Γ. Since Γ is exception-free, every element of $\Phi - \Gamma$ is inheritable in Φ, so Φ is grounded in Γ.   ∎

2.21   General multiple inheritance systems

Far fewer knowledge representation languages permit multiple inheritance with exceptions than use tree-structured inheritance. Exceptions are easy to cope with in tree-structured systems. KLONE is interesting because it does provide multiple inheritance but bars exceptions on philosophical grounds (Brachman, 1985).

In the case of NETL, the general theory developed in this chapter defines what the semantics of inheritance *should* be. The actual semantics was never formally stated, but can be inferred from the published algorithms in (Fahlman, 1979) to be using a shortest-path ordering among competing inferences, with ill effects as outlined in (Fahlman, Touretzky, and van Roggen 1981). My generic system and its inferential distance ordering grew out of an attempt to correctly formalize the intuition underlying Fahlman's work.

3 Predicate Lattices and Formal Semantics

"It is, I guess, in the nature of us all to wonder why the universe appears just the way it does. Why, for example, does it not appear more symmetrical? Well, if you will be kind enough, and patient enough, to bear with me through the argument as it develops itself in this text, you will I think see, even though we begin it as symmetrically as we know how, that it becomes, of its own accord, less and less so as we proceed.

— G. Spencer Brown, *Laws of Form*

3.1 Introduction

Atomic predicates are those such as elephant or gray thing that are represented by nodes in the inheritance graph. In this chapter we will see how the extensions of *complex* predicates, such as "gray elephant," can be constructed from unions, intersections, and complements of the extensions of atomic ones. We will use elementary lattice theory to describe truth values, predicates, and extensions of predicates. The lattice concepts we develop will allow us to give a mathematical semantics to inheritance systems with exceptions.

3.2 Lattices

A lattice is a set with meet and join operators $\wedge$ and $\vee$ satisfying the following identities (MacLane and Birkhoff, 1967):

L1. $x \wedge x = x \vee x = x$ (Idempotent)

L2. $x \wedge y = y \wedge x, \quad x \vee y = y \vee x$ (Commutative)

L3. $x \wedge (y \wedge z) = (x \wedge y) \wedge z,$ (Associative)
$x \vee (y \vee z) = (x \vee y) \vee z$

L4. $x \wedge (x \vee y) = x \vee (x \wedge y) = x$ (Absorption)

A distributive lattice also satisfies L5 and the equivalent L5'.

L5. $x \wedge (y \vee z) = (x \wedge y) \vee (x \wedge z)$ (Distributive)

L5'. $x \vee (y \wedge z) = (x \vee y) \wedge (x \vee z)$

A boolean algebra is a distributive lattice satisfying L6 through L8, with a function
~ (complement) from lattice elements to lattice elements, and elements 0 and 1 such
that 0 and 1 are smallest and largest elements, respectively, under the order in which
$x \leq y$ iff $x \wedge y = x$.

L6. $x \wedge \sim x = 0, \qquad x \vee \sim x = 1$

L7. $\sim(\sim x) = x$

L8. $\sim(x \wedge y) = \sim x \vee \sim y,$

$\sim(x \vee y) = \sim x \wedge \sim y$

In this thesis, all lattices discussed will be finite.

3.3 Powers of a boolean algebra

The set $\{0, 1\}$ is a boolean algebra under the logical operations of conjunction, disjunction, and negation. Figure 3.1 is a picture of this algebra.

1

0

Figure 3.1: The boolean algebra $\{0, 1\}$

Any power of a distributive lattice is itself a distributive lattice, and any power of
a boolean algebra is a boolean algebra. Figure 3.2 shows the boolean algebra $\{0, 1\}$
raised to the third power; the elements of this lattice are n-tuples of 0's and 1's. (In
this example $n = 3$.) The meet, join, and lattice complement operations ($\wedge$, $\vee$, and
$\sim$) on n-tuples are normally defined in terms of the corresponding logical operations on
individual booleans as follows: the meet of two n-tuples is their component-wise logical
conjunction, the join is their component-wise logical disjunction, and the complement
of an n-tuple is its component-wise logical negation.

80

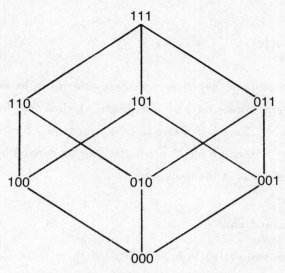

Figure 3.2: The boolean algebra $\{0, 1\}^3$

3.4 Three isomorphic types of lattices

In this chapter we will work with three distinct types of lattices: lattices whose elements are n-tuples of truth values, lattices whose elements are subsets of a universe of n objects, and lattices whose elements are predicates over a domain of n objects. We shall refer to these three types of lattices as truth value lattices, extension lattices, and predicate lattices, respectively. All three types are isomorphic to each other. In later sections we will speak of them informally as if they were the same.

We shall write the truth value lattice for a given value of n as $\{0, 1\}^n$. Given a universe of n elements, the extension lattice consists of the 2^n distinct subsets of this universe, which form a boolean algebra under intersection, union, and set complement that is isomorphic to $\{0, 1\}^n$ under meet, join, and lattice complement. Since each subset defines the extension of some predicate, there are exactly 2^n distinct predicates over this the universe, and these predicates under conjunction, disjunction, and negation form the predicate lattice isomorphic to $\{0, 1\}^n$. We can define the conjunction, disjunction, and negation operations on predicates in the predicate lattice pointwise in terms of logical operations on truth values, as follows:

$$(P_i \land P_j)(x) = P_i(x) \land P_j(x)$$

$$(P_i \lor P_j)(x) = P_i(x) \lor P_j(x).$$

$$(\sim P_i)(x) = \sim P_i(x).$$

Each of the n elements in the universe we have chosen can be assigned a characteristic predicate in the predicate lattice: a predicate which is true for that element and false for all the others. The set of n characteristic predicates forms a *basis set* for the entire predicate lattice, because all 2^n predicates can be derived by meets, joins, and complements from elements of the basis set.

3.5 The three-valued case

We now switch from two-valued logic to a logic of three truth values, T, F, and Q, corresponding to the three signs $+$, $-$, and $\#$ that mark the tokens of our inheritance system. The truth lattice $\{T, F, Q\}$ is shown in figure 3.3. This lattice is distributive

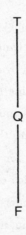

Figure 3.3: A three-valued truth lattice $\{T, F, Q\}$.

because every chain is a distributive lattice, but it is not a boolean algebra because it violates identity L6 when $x = Q$, *viz.*:

Q $\land$ $\sim$Q is Q instead of F, and
Q $\lor$ $\sim$Q is Q instead of T.

Truth tables for the three-valued versions of the conjunction, disjunction, and negation functions appear below.

$\wedge$	T	Q	F
T	T	Q	F
Q	Q	Q	F
F	F	F	F

$\vee$	T	Q	F
T	T	T	T
Q	T	Q	Q
F	T	Q	F

$\sim$	
T	F
Q	Q
F	T

In a three-valued logic, there are 3^n distinct predicates over a domain of n elements, forming a lattice isomorphic to $\{T, F, Q\}^n$. Figure 3.4 shows the lattice $\{T, F, Q\}^2$, and

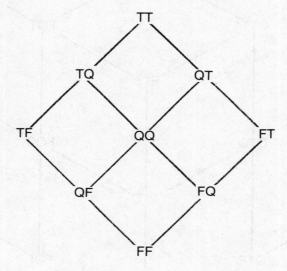

Figure 3.4: The truth lattice $\{T, F, Q\}^2$.

figure 3.5 shows $\{T, F, Q\}^3$. From now on we will identify the lattice of predicates over a universe of n objects with the isomorphic lattice of truth values $\{T, F, Q\}^n$, and speak of the members of $\{T, F, Q\}^n$ as predicates.

We define the meet, join, and complement operations on elements of the predicate lattice pointwise in terms of the conjunction, disjunction, and negation functions on their truth values. That is, for any pair of predicates P_i and P_j in $\{T, F, Q\}^n$, their meet join, and complement are defined at every point by

$$(P_i \wedge P_j)(x) = P_i(x) \wedge P_j(x)$$

83

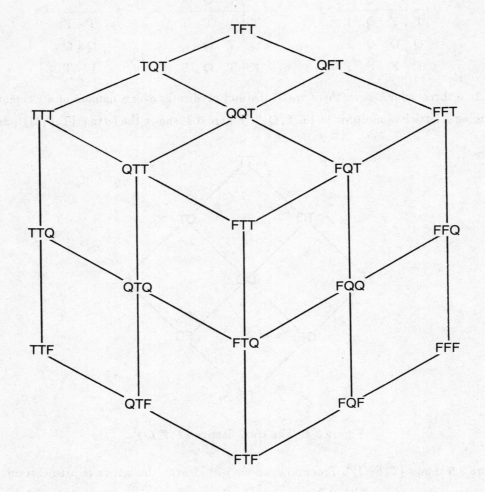

Figure 3.5: The truth lattice $\{T, F, Q\}^3$ drawn as an opaque cube.

84

$$(P_i \lor P_j)(x) = P_i(x) \lor P_j(x)$$

$$(\sim P_i)(x) = \sim P_i(x).$$

3.6   Extensions of three-valued predicates

The extensions of three-valued predicates must be defined slightly differently than in the two-valued case. We will write the extension of a three-valued predicate P_i as a triple of sets $\langle T_i, F_i, Q_i \rangle$ forming a three-way partition of the domain. An element is in the set T_i iff predicate P_i has value T for it; similarly it is in F_i if the predicate has value F and in Q_i if it has value Q. The set of 3^n distinct extensions in a universe of n elements forms an extension lattice that is obviously isomorphic to the truth value lattice $\{T, F, Q\}^n$.

Note that $T_i \cup F_i \cup Q_i = \Pi$ for all P_i in the predicate lattice. There are maximal and minimal elements of this lattice, denoted by I and O, respectively, such that for every P_i, $(P_i \land I) = (P_i \lor O) = P_i$. Let Π denote the universe set of which elements of the extension lattice are partitionings. The extension of the maximal predicate, I, is $\langle \Pi, \emptyset, \emptyset \rangle$; that of the minimal predicate, O, is $\langle \emptyset, \Pi, \emptyset \rangle$. For symmetry, let U denote the maximally ambiguous predicate, whose extension is $\langle \emptyset, \emptyset, \Pi \rangle$.

We can define meet, join, and lattice complement operations on elements of the extension lattice in terms of intersections, unions, and set complements on their component T, F, and Q sets, as follows:

$$\langle T_i, F_i, Q_i \rangle \land \langle T_j, F_j, Q_j \rangle \quad = \quad \langle T_i \cap T_j, \quad F_i \cup F_j, \quad \Pi - (T_i \cap T_j) - (F_i \cup F_j) \rangle$$

$$\langle T_i, F_i, Q_i \rangle \lor \langle T_j, F_j, Q_j \rangle \quad = \quad \langle T_i \cup T_j, \quad F_i \cap F_j, \quad \Pi - (T_i \cup T_j) - (F_i \cap F_j) \rangle$$

$$\sim \langle T_i, F_i, Q_i \rangle \quad = \quad \langle F_i, T_i, Q_i \rangle$$

The meet, join, and complement operations on extensions of three-valued predicates are clearly isomorphic to meet, join, and complement operations on the predicates themselves. The above equations will play a significant role in chapter 4, where lattice operations on predicates will be implemented in terms of set operations on the component T, F, and Q sets of their extensions. Set operations can be performed extremely rapidly on parallel marker propagation machines.

3.7 The universe of the inheritance graph

We now move from the general case of a universe with n elements to the case where the elements are nodes in an inheritance graph. Let Π be the set of nodes of an inheritance graph. There are $3^{|\Pi|}$ distinct three-valued predicates in this universe, forming a lattice $\{T, F, Q\}^\Pi$ which we will name $\mathcal{L}_\Pi$. The nodes of an inheritance graph are real-world individuals and real-world predicates. To prevent confusion, we will refer to the elements of $\mathcal{L}_\Pi$ as *lattice* predicates.

The domain of real-world predicates is the set of real-world individuals, *i.e.* the elements of Π that are individuals. Thus from $\langle +clyde, +mammal \rangle$ we can infer that the real-world mammal predicate is true of Clyde, and express this fact in logical notation by writing Mammal(clyde). But from $\langle +elephant, +mammal \rangle$ it makes no sense to infer that the mammal predicate is true of elephant, because elephant is not an individual. Lattice predicates, on the other hand, range over all the objects in Π, so if P_i is a lattice predicate representing the real-world mammal predicate, writing $P_i(elephant)$ is as correct as writing $P_i(clyde)$. Not all lattice predicates represent real-world predicates. Those that do are called E-predicates, for reasons that will become clear shortly. Call the lattice predicate representing the real-world mammal predicate E_{mammal}. Then "Clyde is a mammal" is expressed by $E_{mammal}(clyde)$, and "elephants are mammals" by $E_{mammal}(elephant)$. In the next section we will formally define the association between E-predicates and their corresponding real-world predicates.

3.8 E-predicates

For every real-world predicate y and consistent expansion Φ there exists a unique lattice predicate which we will call E_y. The lattice predicate E_y has the same value as y for all individuals in Π, but E_y is defined over the predicates in Π as well. Thus, the extension of E_y is in agreement with that of y for all individual nodes in their common domain. (The E in E-predicate stands for extension, for this reason.) We formally define the association between the lattice predicate E_y and the real-world predicate y by showing how the extension of E_y, written $\langle T_y, F_y, Q_y \rangle$, is derived from the conclusion set $C(\Phi)$:

$$T_y = \{x \mid x = y \text{ or } \langle +x, +y \rangle \in C(\Phi)\}$$

$$F_y = \{x \mid \langle +x, -y \rangle \in C(\Phi)\}$$

$$Q_y = \Pi - (T_y \cup F_y)$$

We make $E_y(y)$ be T by definition because "the typical y" should certainly satisfy the real-world y predicate, *e.g.* the typical elephant is certainly an elephant. Note that if $E_y(y)$ were F or Q, that would imply that Γ contained $\langle +y, -y \rangle$ or $\langle +y, \#y \rangle$, in which case Γ would be inconsistent. Defining $E_y(y)$ to be T results in an important property for the meets of E-predicates which we illustrate by example. If "elephants are mammals," then $E_{elephant} \wedge E_{mammal}$ will contain *elephant* in the T part of its extension, meaning the set of things which are both mammals and elephants includes the typical elephant. And of course, it should.

The correspondence between E-predicates and elements of Π can be extended to include real-world individuals. Each individual $y \in \Pi$ has an E-predicate whose extension is $\langle \{y\}, \emptyset, \Pi - \{y\} \rangle$. Real-world individuals and predicates are treated identically as far as lattice predicates are concerned. The distinction between the two types of real-world object in our formal system may therefore seem pointless, but we maintain it because it is intuitively helpful. It might matter in elaborations of the formal system.

Theorem 3.1 If Γ is IS-A acyclic, then for every $x, y \in \Pi$, $E_x = E_y$ iff $x = y$.

Proof Suppose $E_x = E_y$. Then $E_x(y) = $ T and $E_y(x) = $ T. If $x \neq y$ then Φ must contain two sequences $\langle +x, +z_1, \ldots, +z_n, +y \rangle$ and $\langle +y, +w_1, \ldots, +w_m, +x \rangle$, so $x \prec y$ and $y \prec x$. This violates the assumption that Γ is i.a. ∎

3.9 Constructability of lattice predicates

Our inheritance language affords us a set of nodes, each of which is assigned a corresponding E-predicate by an expansion. We define $E_\Pi(\Phi)$ to be the set of E-predicates determined by the expansion Φ, *i.e.* the set $\{E_y \mid y \in \Pi\}$. Where the expansion Φ is clear, we will simply write E_Π. Note that $\bigvee_{y \in \Pi} E_y = I$ and therefore $\bigwedge_{y \in \Pi} \sim E_y = O$. Consider the lattice $L(E_\Pi)$ formed by the closure of meet, join, and complement operations with E_Π as the basis set. This is the lattice of *constructable* predicates determined by the expansion. As noted above, it contains the maximal and minimal elements I and O. $L(E_\Pi)$ is clearly a sublattice of $\mathcal{L}_\Pi$, the complete predicate lattice, since all its elements are elements of $\mathcal{L}_\Pi$ and it is closed under meet, join, and complement.

Figures 3.6 through 3.12 each show a three-node inheritance graph with a unique grounded expansion and the corresponding constructable sublattice $L(E_\Pi)$. In the lattice portion of the figure, each oval containing a triple of truth values xyz denotes a lattice predicate P_i defined by $P_i(a) = x$, $P_i(b) = y$, and $P_i(c) = z$. For example, the oval labeled TQQ in figure 3.6 denotes the lattice predicate E_A, with extension $\langle \{a\}, \emptyset, \{b, c\}\rangle$.

Figure 3.11 is a particularly interesting case because the network contains an inheritance exception. The asymmetry of the inheritance network is reflected in the asymmetry of its expansion's constructable sublattice.

Note that the lattices of *constructable* predicates in figures 3.6 through 3.12 contain between 6 and 18 elements. The predicate lattice $\mathcal{L}_\Pi$ for a universe of three nodes contains 3^3, or 27 elements, as shown in figure 3.5. Most graphs have expansions whose constructable sublattices are incomplete. However, figure 3.13 presents two graphs whose expansions have *complete* constructable lattices, *i.e.* the lattice of constructable predicates is $\mathcal{L}_\Pi$.

3.10 A mathematical semantics for inheritance systems

We are now in a position to give a formal semantics for inheritance systems in terms of lattices of predicates. (A formal semantics for inheritance in the Omega system was reported by Attardi and Simi (1982). However, their inheritance system did not permit exceptions; the problem is entirely different when exceptions are present.)

Before proceeding with my own definitions, I should explain why the set-theoretic semantics ordinarily used with first order logic does not apply to nonmonotonic systems, which permit exceptions. Consider the statement "elephants are gray." In first order logic this becomes

$$(\forall x) \quad \text{Elephant}(x) \rightarrow \text{GrayThing}(x).$$

The interpretation we give to this statement is that the extension of the elephant predicate is a subset of the extension of the gray-thing predicate. But in an inheritance system that permits exceptions, this interpretation may not be valid. In inheritance reasoning, "elephants are gray" can be true even when *none* of the instances of elephant in the knowledge base are gray. (Perhaps we are only interested in recording non-gray ones.)

The meanings ordinarily given to statements in first order logic are stated in terms of the extensions of predicates, rather than in terms of their intensions. But in an inheritance system, "elephants are gray" says nothing concrete about the extensions of elephant or gray. It is a statement about the intension of the elephant predicate, not about its extension. It tells us that if we know an object is an elephant then we should conclude that it is gray *unless* we have information to the contrary. Since the statement is intensional, its meaning cannot be defined in terms of the extension of the elephant predicate.

Besides dealing with exceptions, another problem that must be solved if we are to have an acceptable semantics for inheritance systems is that of representing the properties of classes as well as of individuals. For example, if we have "birds can fly" and "canaries are birds" we should conclude that "canaries can fly" is true and "canaries cannot fly" is false, even in worlds where there are no known instances of canaries, or in which all known canaries happen to have broken wings and therefore cannot fly. And in answer to the query "who can fly" we should get back not only individual birds such as Tweetie, but also classes such as canary — again, even if there are no known instances of the class. An advantage of phrasing meanings in terms of E-predicates, then, is that we obtain a uniform representation for propositions such as "Tweetie can fly" and intensional statements such as "canaries can fly," which is in keeping with the uniform representation for those statements in the inheritance system given in chapter 2.

We are now ready to define formal interpretations for the following types of objects:

- An individual sequence $\sigma \in \Sigma$.

- A set of sequences $S \subseteq \Sigma$.

- A *circumscribed* set of sequences $S \subseteq \Sigma$. By circumscribed, we mean mean that S is to be interpreted as a complete set of axioms. No additional assertions may be added to S except those that are grounded in S.

- An inheritance graph $\Gamma \subseteq \Theta \times \Theta \subseteq \Sigma$.

- A node $x \in \Pi$, representing a predicate or an individual in the inheritance graph.

First, for every set of sequences $S \subseteq \Sigma$, let $Inh(S)$ denote the set of all sequences inheritable in S. It doesn't matter whether these sequences are contained in S or not.

Now we define the formal interpretation of a sequence σ, denoted $I(\sigma)$, to be the set of all sets in which $\sigma = \langle x_1, \ldots, x_n \rangle$ has been extended to $\langle x_1, \ldots, x_n, y \rangle$ in all possible ways consistent with inheritability. Formally,

$$I(\sigma) = \{ S \subseteq \Sigma \mid (\forall y) \quad \langle x_1, \ldots, x_n, y \rangle \in Inh(S) \rightarrow \langle x_1, \ldots, x_n, y \rangle \in] \} \,.$$

We can extend the interpretation of single sequences in a straightforward way to provide a natural interpretation of a set of sequences. If S is any set of sequences, then

$$I(S) = \bigcap_{\sigma \in S} I(\sigma).$$

S is not circumscribed in this definition, and it need not be closed under inheritance. Note that S is closed under inheritance iff $S \in I(S)$. Note also that $I(S)$ contains all closed supersets of S.

Next consider the circumscribed interpretation of S, denoted $I(\lceil S \rceil)$. When we circumscribe a set, this means the set is to be interpreted as a complete set of axioms, thereby eliminating ungrounded supersets from its interpretation. See (McCarthy, 1980) for a discussion of circumscription as a form of nonmonotonic reasoning. We define $I(\lceil S \rceil)$ as follows:

$$I(\lceil S \rceil) = \{ \Phi \mid \Phi \in I(S) \text{ and } \Phi \text{ is grounded in } S \} \,.$$

Thus for inheritance graphs Γ, which we view as circumscribed sets, each element of $I(\lceil \Gamma \rceil)$ is a grounded expansion of Γ and represents one possible interpretation of the inheritance graph.

We define $I(x, \Gamma)$, the interpretation of a node x in an inheritance graph Γ, to be a *function* whose input is a grounded expansion from the set $I(\lceil \Gamma \rceil)$ and whose output is an E-predicate. The interpretation of a node x with respect to a particular grounded expansion Φ is given by $I(x, \Gamma)(\Phi)$. We define the value of $I(x, \Gamma)(\Phi)$ to be E_x as determined by Φ if Γ is consistent, O otherwise.

Recall that $E_\Pi(\Phi)$ is the set of all E-predicates determined by the grounded expansion Φ. Let $L(E_\Pi(\Phi))$ denote the predicate sublattice constructable from $E_\Pi(\Phi)$. For each inheritance graph Γ there is a set of constructable sublattices $\{ L(E_\Pi(\Phi)) \mid \Phi \in I(\lceil \Gamma \rceil) \}$ that correspond to the set of grounded expansions of that graph. Some examples of constructable sublattices were given in figures 3.6 through 3.12. Thus we see that an inheritance graph determines a set of sublattices, and the meaning we give to a node in the graph is a mapping into an element of each lattice, as determined by the graph's grounded expansions.

This completes our specification of the predicate lattice model of inheritance systems. Now we have an exact formal interpretation of inheritance that is independent of any vague notions of typicality, prototypes, and so forth, yet powerful enough to represent intensional properties of classes.

3.11 A-predicates as the duals of E-predicates

In chapter 4 we will study a machine that can compute the extensions of certain lattice predicates extremely rapidly. One such class of predicates are the E-predicates. Computing their extensions allows us to answer questions such as "what things are elephants" or "what things are gray." Another important type of question is "what are the properties of elephants." In order to phrase questions of this type in terms of extensions of lattice predicates, we define another class of distinguished lattice predicates known as A-predicates (A for abstraction) that are the duals of E-predicates (E for extension.) For any $x \in \Pi$ we define A_x, the A-predicate corresponding to node x, with respect to a particular expansion Π by noting that there is a set of E-predicates that are true of x, a set that are false, and a set that are inconclusive. We write the extension of A_x as a triple $\langle T^x, F^x, Q^x \rangle$ where the sets T^x, F^x, and Q^x are defined as:

$$T^x = \{y \mid E_y(x) = \mathrm{T}\}$$

$$F^x = \{y \mid E_y(x) = \mathrm{F}\}$$

$$Q^x = \{y \mid E_y(x) = \mathrm{Q}\}$$

Note that $T^x \cup F^x \cup Q^x = \Pi$. We define the set A_Π to be $\{A_x \mid x \in \Pi\}$.

Theorem 3.2 For all $x, y \in \Pi$, $E_y(x) = A_x(y)$.

Proof   Let $\langle T^x, F^x, Q^x \rangle$ denote the extension of A_x. If $E_y(x) = \mathrm{T}$ then $y \in T^x$, so $A_x(y) = \mathrm{T}$. Similarly, if $E_y(x) = \mathrm{F}$ then $y \in F^x$ so $A_x(y) = \mathrm{F}$, and if $E_y(x) = \mathrm{Q}$ then $y \in Q^x$ so $A_x(y) = \mathrm{Q}$. Therefore, $E_y(x) = A_x(y)$. ∎

The predicate A_x is called the *abstraction predicate* for node x because its extension lists the *properties* of x, e.g. if "elephants are gray" then $A_{elephant}$ includes gray in the T part of its extension. By computing the extension of an abstraction predicate we can obtain a list of all the properties of an individual, or of a pseudo-individual such as the typical elephant. Note that the extension of the real-world gray thing predicate contains only individuals who are gray, such as Clyde. The extension of the three-valued lattice predicate $E_{gray.thing}$ contains all of Clyde, elephant and gray thing in its T part. In the T-part of $A_{elephant}$ we find predicates such as gray thing, mammal, quadruped and of course, elephant.

Theorem 3.3 For every expansion of an IS-A acyclic Γ, if $y \in T_x - \{x\}$ and $z \in T^x - \{x\}$, then $y \prec x \prec z$.

Proof If $y \in T_x - \{x\}$ then $\langle +x, +y \rangle \in C(\Phi)$, which means the expansion contains a sequence of form $\langle +x, \ldots, +y \rangle$. Therefore $x \prec y$. If $z \in T^x - \{x\}$ then $x \in T_z$, so by the preceding argument $x \prec z$. ∎

In network terms, the relationship between E-predicates and A-predicates is that T_x, the T part of the extension of E_x, consists of x and nodes below x in the IS-A hierarchy, while T^x, the T part of the extension of A_x, consists of nodes above x in the hierarchy.

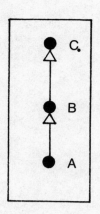

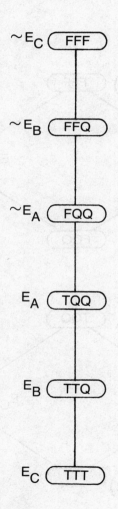

~E_C FFF

~E_B FFQ

~E_A FQQ

E_A TQQ

E_B TTQ

E_C TTT

Figure 3.6: $L(E_\Pi)$ contains 6 elements.

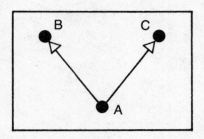

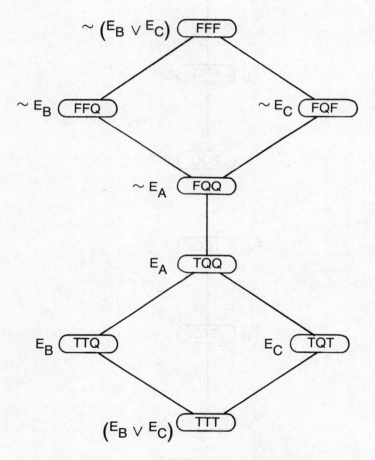

Figure 3.7: $L(E_\Pi)$ contains 8 elements.

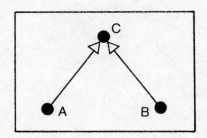

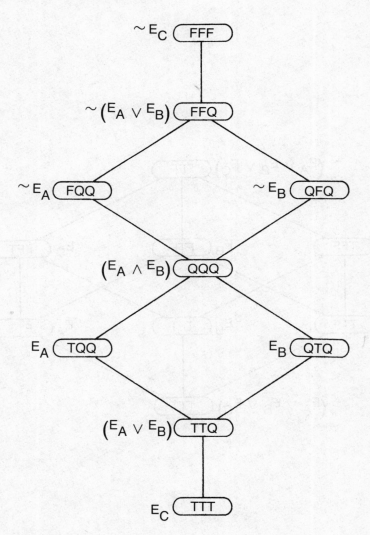

Figure 3.8: $L(E_\Pi)$ contains 9 elements.

95

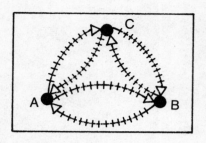

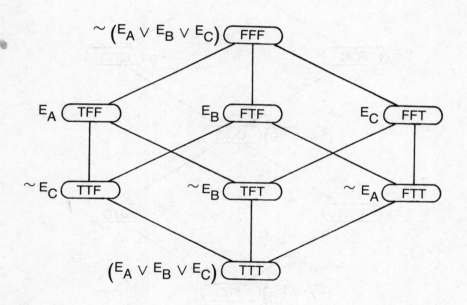

Figure 3.9: $L(E_\Pi)$ is the powerset lattice, containing $2^3 = 8$ elements.

96

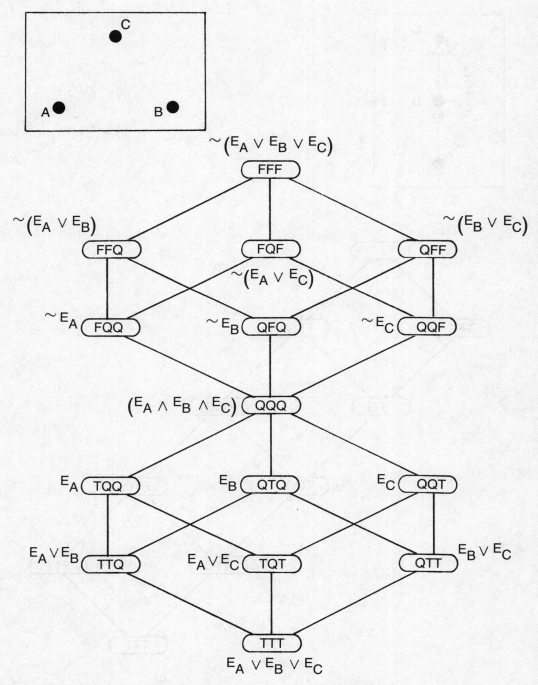

Figure 3.10: $L(E_\Pi)$ contains 15 elements.

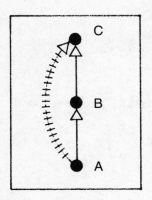

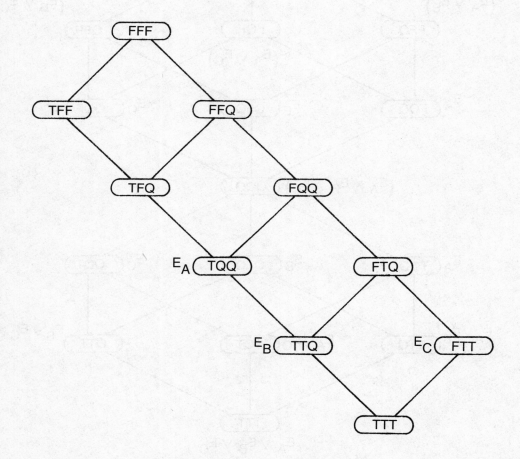

Figure 3.11: $L(E_\Pi)$ contains 10 elements.

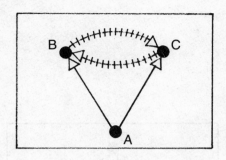

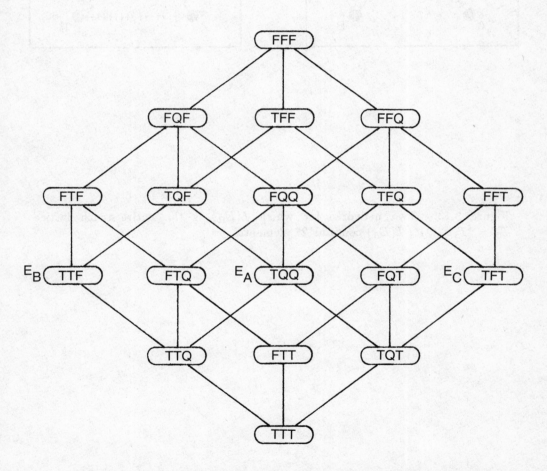

Figure 3.12: $L(E_\Pi)$ is a $3 \times 3 \times 2$ array containing 18 elements.

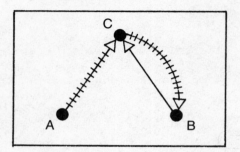 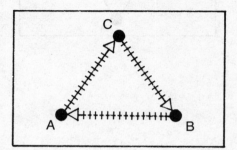

Figure 3.13: Two networks for which $L(E_\Pi)$ is the entire truth lattice $\mathcal{L}_\Pi = \{T, F, Q\}^3$. $L(E_\Pi)$ contains 27 elements.

4 Parallel Marker Propagation Machines

"The human mind can do many remarkable things. Of these, perhaps the most remarkable is the mind's ability to store a huge quantity and variety of knowledge about the world, and to locate and retrieve whatever it needs from this storehouse at the proper time. This retrieval is very quick, very flexible, and in most cases seems almost effortless. If we are ever to create an artificial intelligence with human-like abilities, we will have to endow it with a comparable knowledge-handling facility."

— Scott Fahlman, *NETL: A System for Representing and Using Real-World Knowledge*

4.1 Introduction

Marker propagation was first introduced in AI through Ross Quillian's work on semantic memory (Quillian, 1968). Later, Scott Fahlman demonstrated that marker propagation could be used for efficient inheritance reasoning (Fahlman, 1979). The technique has recently been applied to word sense and case slot disambiguation (Hirst and Charniak, 1982), understanding noun compounds (MacDonald, 1982), and taxonomic classification (Woods, 1978).

In this chapter we analyze a massively parallel computer architecture known as a parallel marker propagation machine (PMPM), invented by Fahlman, that appears well-suited for inheritance reasoning. Inheritance networks are represented on a PMPM by assigning a separate hardware element to each node and each link. These elements have a tiny bit of local memory used to represent markers, and some connections to other elements through which the markers propagate. Elements receive instructions over a global command bus and execute them in parallel. My treatment of parallel marker propagation-based inheritance differs from Fahlman's in several ways:

- I provide a theoretical context for analyzing PMPM algorithms by relating the states of the parallel marker propagation machine to the elements of $\mathcal{L}_\Pi$, the predicate lattice defined in chapter 3. We will see how the formal analysis of inheritance in chapter 2 and the mathematical semantics given to inheritance systems in chapter 3 serve to define "correctness" for marker propagation algorithms.

- My algorithms are provably correct. In contrast, earlier PMPM algorithms developed for NETL, from which my own algorithms are derived, could produce incorrect results when run on even slightly complex networks (Fahlman, Touretzky, and van Roggen, 1981).

- My algorithms are more sophisticated than earlier ones. They implement a three-valued logic that was hinted at but not thoroughly explored in NETL. They also obey an important symmetry law that was not upheld by the algorithms in (Fahlman, 1979).

4.2 Graph coloring

In an abstract sense, marker propagation is just graph coloring. Suppose we have three colors T, F, and Q, and an inheritance graph with a set of nodes Π. If each node is colored with exactly one of T, F, and Q, then the colored graph as a whole represents the extension of some predicate $P_i \in \mathcal{L}_\Pi$ such that $P_i(x) = c$ iff node x is colored with color c. If we have n distinct triples of colors and each node is colored with exactly one color from each triple, then the graph represents the extensions of n lattice predicates simultaneously.

In this chapter we will perform two very different kinds of operations by coloring the nodes of an inheritance graph. First, we will compute meets, joins, and complements of lattice predicates by coloring the graph with additional triples based on the colors it presently bears. Second, when we want to reconstruct the extension of a particular E-predicate or A-predicate, we will color the graph by following certain rules based on the graph's topology rather than the colors it bears at the moment.

4.3 Marker bits

Marker bits are boolean values that correspond to colors. The colors of the ith triple, T_i, F_i, and Q_i, are represented by marker bits called TM_i, FM_i, and QM_i respectively. Each node of the graph has its own set of marker bits. Saying that node x has its TM_i bit on is equivalent to saying that x bears color T_i.

102

The hardware details of a PMPM are discussed in (Fahlman, 1980), but let me explain briefly why hardware considerations force us to separate the idea of a marker triple from that of a particular predicate's extension. A graph of N nodes requires 3×3^N marker bits to represent the extension of every predicate in $\mathcal{L}_\Pi$ simultaneously. To represent the extensions of the E-predicates alone takes $3N$ bits; for a 100,000 node graph, that translates to 37 kilobytes of memory per node. Since we are interested in representing inheritance graphs in hardware, using currently available technology, we can afford far fewer than N bits of storage per node if we are to have a large number of nodes. Therefore we cannot assign each lattice predicate, or even each E-predicate, its own triple of markers. Instead we must use markers as a form of temporary physical storage in which any extension can be represented, much as a single page of physical memory can represent any page of virtual memory on a conventional computer.

4.4 A language for PMPM algorithms

Throughout this chapter we will write PMPM algorithms in a high level marker propagation language. This language can be compiled into actual commands to a PMPM and its associated control computer, or it can be executed interpretively by a PMPM simulator. I have written such a simulator in MacLisp. Scott Fahlman and Walter van Roggen have also written simulators, but their programs interpret a lower level language than mine — one that is less abstract but more faithful to the actual hardware. The details of my abstract PMPM language are presented below.

The simplest statement in the PMPM language is an unconditional action. Actions set or clear marker bits. Although marker bits are usually referenced as TM_i, FM_i, or QM_i, we will write M_i to refer to a marker bit in general, independent of the kind of color it represents. The notation for the two basic actions a node can perform on marker bits is given in table 4.1.

$\textbf{set}[M_1, \ldots, M_n]$ set marker bits M_1 through M_n

$\textbf{clear}[M_1, \ldots, M_n]$ clear marker bits M_1 through M_n

Table 4.1: The basic actions a node can perform on marker bits.

Actions on marker bits are performed by all nodes simultaneously. For example, the action **clear**$[TM_1]$ causes all nodes to clear their TM_1 bit, *i.e.* to erase color T_1 from themselves.

If we perform indiscriminate actions on marker bits then a coloring may not represent the extension of any predicate. For instance, if some node in the graph has none of TM_i, FM_i, or QM_i set, or if it has more than one of them set, then the ith marker triple does not represent the extension of any predicate in $\mathcal{L}_\Pi$ and so is not a valid coloring.

The next statement type in our PMPM command language is the conditional action. Its syntax is:

$$cond_1, \ldots, cond_n \implies action_1, \ldots, action_m$$

The conditions are treated conjunctively. If all are true, each of the actions will be executed. Conditions refer to bits that are on or off at a node, or to the unique name (serial number) assigned to a node. The basic conditions are shown in table 4.2. The use of node serial numbers is discussed later in the chapter.

on$[M_1, \ldots, M_n]$	each of M_1 through M_n is on
off$[M_1, \ldots, M_n]$	each of M_1 through M_n is off
any-on$[M_1, \ldots, M_n]$	at least one of M_1 through M_n is on
any-off$[M_1, \ldots, M_n]$	at least one of M_1 through M_n is off
name$[x]$	node has name (serial number) x

Table 4.2: Conditions for which a node can test.

Conditional actions, like unconditional ones, are processed by all nodes simultaneously. If the conditions are true at a given node, the actions will be performed by that node. If the conditions are false at a node, the node will ignore the actions. The following conditional command causes all nodes with both TM_1 and TM_2 on to turn on TM_3. Nodes that don't have both TM_1 and TM_2 on will leave their TM_3 bit unaffected.

$$\textbf{on}[TM_1, TM_2] \implies \textbf{set}[TM_3]$$

We now have a rich enough set of commands to compute meets, joins, and complements of predicates by manipulating marker bits. For example, let us assume the ith

104

```
procedure complement(i, j: marker-triple) = begin
    clear[TM_j, FM_j, QM_j];
    on[TM_i]  ⟹  set[FM_j];
    on[FM_i]  ⟹  set[TM_j];
    on[QM_i]  ⟹  set[QM_j]
end
```

Figure 4.1: A procedure for computing the complement of a lattice predicate.

marker triple represents the extension of some lattice predicate; we wish to represent its complement with the jth marker triple. Recall from chapter 3 the definition of the complement of an extension:

$$\sim \langle T_i, F_i, Q_i \rangle \quad = \quad \langle F_i, T_i, Q_i \rangle.$$

Figure 4.1 shows the procedure for computing complements on a PMPM. Procedure **complement** takes marker triples i and j as input. Upon entry i is assumed to represent the extension of a lattice predicate; upon exit, j will represent its complement. (We require that $i \neq j$.)

Next, recall the definition of the meet of two extensions:

$$\langle T_i, F_i, Q_i \rangle \wedge \langle T_j, F_j, Q_j \rangle = \langle T_i \cap T_j,\ F_i \cup F_j,\ \Pi - ((T_i \cap T_j) \cup F_i \cup F_j) \rangle.$$

To compute the meet of the extensions represented by two marker triples, representing the result with a third marker triple, we use the procedure shown in figure 4.2. The procedure for computing joins is left as an exercise for the reader.

The three procedures **meet, join,** and **complement** allow us to find the extension of any predicate constructable from a given basis set. We are interested particularly in the basis set E_Π, which for a graph of N nodes contains N members. Yet when N is large, hardware constraints prevent us from representing the extensions of all N predicates at once. This brings us to the second type of marker operation we will consider in this chapter: those that allow us to reconstruct the extensions of E-predicates and A-predicates as needed. To do so we must first define some additional PMPM commands.

```
procedure meet(i, j, k: marker-triple) = begin
  clear[TM_k, FM_k, QM_k];
  on[TM_i, TM_j]  ⟹  set[TM_k];
  any-on[FM_i, FM_j]  ⟹  set[FM_k];
  off[TM_k, FM_k]  ⟹  set[QM_k]
end
```

Figure 4.2: A procedure for computing the meet of two lattice predicates.

4.5 Link commands

The commands covered in the previous section are executed only by nodes. Our PMPM language also contains link commands, control commands (for iteration), and i/o commands for reporting results (which we omit from this discussion.)

Since each link has a node at its head and a node at its tail, we will allow the actions performed by a link to affect the markers of either its head or tail nodes, or both. The basic link actions are shown in table 4.3.

set-head$[M_1, \ldots, M_n]$	turn on head node's bits M_1 through M_n
clear-head$[M_1, \ldots, M_n]$	turn off head node's bits M_1 through M_n
set-tail$[M_1, \ldots, M_n]$	turn on tail node's bits M_1 through M_n
clear-tail$[M_1, \ldots, M_n]$	turn off tail node's bits M_1 through M_n

Table 4.3: Actions that a link can perform.

For example, the command **set-tail**$[TM_2]$ causes every link to turn on bit TM_2 of its tail node. In Fahlman's original description of a PMPM, links had their own marker bits and could set and clear their own bits as well as those of their head and tail nodes. We will not make use of this feature here.

Conditional link commands, shown in table 4.4, have the same syntax as conditional node commands, but they use a different set of conditions. We need conditions to test the marker bits of the link's head and tail nodes. We also need a condition to test the type of a link, since most commands will be intended for just certain types of links.

106

link-type$[t_1, \ldots, t_n]$	link is of one of the specified types (such as IS-A, IS-NOT-A, or NO-CONCLUSION)
on-head$[M_1, \ldots, M_n]$	all of the head node's markers $M_1, \ldots, M_n$ are on
off-head$[M_1, \ldots, M_n]$	all of the head node's markers $M_1, \ldots, M_n$ are off
any-on-head$[M_1, \ldots, M_n]$	at least one of head node's $M_1, \ldots, M_n$ is on
any-off-head$[M_1, \ldots, M_n]$	at least one of head node's $M_1, \ldots, M_n$ is off
on-tail$[M_1, \ldots, M_n]$	all of the tail node's markers $M_1, \ldots, M_n$ are on
off-tail$[M_1, \ldots, M_n]$	all of the tail node's markers $M_1, \ldots, M_n$ are off
any-on-tail$[M_1, \ldots, M_n]$	at least one of tail node's $M_1, \ldots, M_n$ is on
any-off-tail$[M_1, \ldots, M_n]$	at least one of tail node's $M_1, \ldots, M_n$ is off

Table 4.4: Conditions for which a link can test.

Each node in the graph has a unique name. At the hardware level the names are actually serial numbers — the Clyde node might have serial number 10347. If we specify a node by name in a conditional command, then the actions associated with that command will be executed by the named node only, not by any other node.

Figure 4.3 shows a procedure that uses a combination of link and node commands to test whether an inheritance graph is consistent with respect to a given node x. (It is not possible to check the consistency of the whole graph on a PMPM, except node-by-node.) If Γ contains any pair of sequences of form $\langle x, y \rangle$, $\langle x, y' \rangle$ or a single sequence of form $\langle x, x' \rangle$, then Γ is inconsistent with respect to x. Procedure consistency-check takes as input a node name, the number i of a marker triple it can use for temporary storage, and the number j of a single marker M_j it should use to mark its result. M_j will be placed on all nodes y such that x is inconsistent with respect to y. If Γ contains $\langle x, x' \rangle$ then x itself will be marked with M_j.

4.6 Transitive closures

PMPM's are especially good at computing transitive closures rapidly. They can do so in time proportional to the depth of the graph, independent of both the number of nodes

```
procedure consistency-check(x: node; i: marker-triple; j: marker) =
   begin
   clear[TM_i, FM_i, QM_i];
   name[x]  ⟹  set[TM_i];
   link-type[IS-A], on-tail[TM_i]  ⟹  set-head[TM_i];
   link-type[IS-NOT-A], on-tail[TM_i]  ⟹  set-head[FM_i];
   link-type[NO-CONCLUSION], on-tail[TM_i]  ⟹  set-head[QM_i];
   clear[M_j];
   on[TM_i, FM_i]  ⟹  set[M_j];
   on[FM_i, QM_i]  ⟹  set[M_j];
   on[QM_i, TM_i]  ⟹  set[M_j]
   end
```

Figure 4.3: A procedure that checks the network for consistency with respect to a given node x.

in the graph and the number of links emanating from each node. Transitive closure is an iterative operation: at each step, markers are propagated in parallel across all eligible links. The iterative construct in our PMPM command language is called **loop**. It has the following syntax:

loop
   body
endloop

Upon entering the loop, the commands in the body are executed once. If at least one conditional command executes successfully by time the end of the body is reached, *i.e.* if the condition part of at least one conditional command is satisfied by at least one node, then the loop repeats. Otherwise the loop terminates. The example procedure in figure 4.4 uses the **loop** construct to compute the transitive closure of link type t starting from node x. The result is represented by a single marker M_i.

Note that the command in the loop body that propagates marker M_i checks that the head node of a link has M_i off before the link is allowed to turn it on. This is so that when the transitive closure is complete, meaning all nodes that should have their M_i bit set do have it set, the condition part of the command in the body will be false for all nodes, so the loop will terminate.

108

```
procedure transclose(x: node; t: link-type; i: marker) = begin
  clear[M_i];
  name[x]  ⟹  set[M_i];
  loop
    link-type[t], on-tail[M_i], off-head[M_i]  ⟹  set-head[M_i]
  endloop
end
```

Figure 4.4: A procedure for computing the transitive closure of a link from a given starting node x.

Figure 4.5 shows how the **transclose** procedure would compute a transitive closure over a set of IS-A links in parallel. Assuming the bottom node of the tree was the input to the procedure, this node would first be marked with marker M_0. Then, upon entering the loop body, the marker would propagate up to two other nodes. The next time through the loop, it would propagate up to four more nodes, all in parallel. After that no new nodes are marked, so the loop would terminate. It is important to note that the running time of this algorithm depends only on the depth of the inheritance graph, not on the number of nodes, the number of links, or the graph's maximum fanout.

4.7 Reconstructing the extensions of predicates

Since a PMPM has only a small number of marker bits per node, there aren't nearly enough triples of markers to represent the extension of every predicate in E_Π or A_Π simultaneously, much less all those in $\mathcal{L}_\Pi$. Instead we must reconstruct extensions as needed and discard them (*i.e.* reuse the marker triples) afterwards. In the case of E-predicates and A-predicates, a PMPM's ability to perform fast transitive closures provides an efficient means of reconstruction. In both cases we use modified transitive closure procedures over IS-A links. Modifications are necessary because in order to construct the full extension we must handle IS-NOT-A and NO-CONCLUSION links as well as IS-A links, and because in the presence of exceptions, IS-A isn't strictly transitive. The procedure for reconstructing A-predicates is the simpler of the two, so we shall consider it first. It is called *upscan* because it propagates markers upward along the IS-A hierarchy. The algorithm in figure 4.6 is essentially the same as Fahlman's.

109

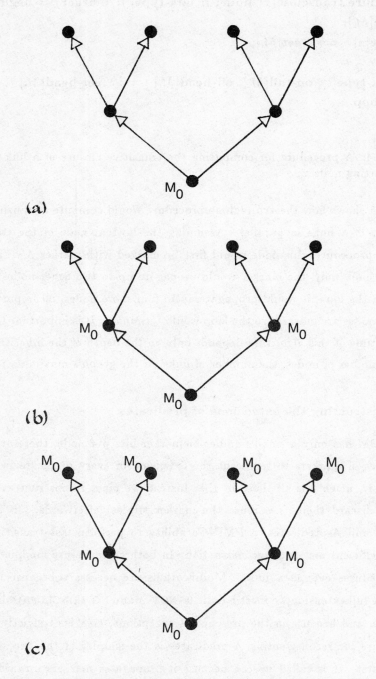

(a)

(b)

(c)

Figure 4.5: Computing a transitive closure over a set of IS-A links.

```
procedure upscan(x: node; i: marker-triple) = begin
  clear[TM_i, FM_i, QM_i];
  name[x]  ⟹  set[TM_i];
  loop
    link-type[IS-A],
      on-tail[TM_i], off-head[TM_i, FM_i, QM_i]  ⟹  set-head[TM_i];
    link-type[IS-NOT-A],
      on-tail[TM_i], off-head[TM_i, FM_i, QM_i]  ⟹  set-head[FM_i];
    link-type[NO-CONCLUSION],
      on-tail[TM_i], off-head[TM_i, FM_i, QM_i]  ⟹  set-head[QM_i]
  endloop;
  off[TM_i, FM_i, QM_i]  ⟹  set[QM_i]
end
```

Figure 4.6: The upscan algorithm.

Figure 4.7 shows the familiar description of Clyde the royal elephant. An upscan of the Clyde node using marker triple number 1 is shown in figure 4.8. First, Clyde is marked with TM_1 (figure 4.8a.) On the first pass through the loop, TM_1 propagates up to royal elephant (figure 4.8b.) On the second pass through the loop, TM_1 propagates to elephant, and the IS-NOT-A link from royal elephant to gray thing causes gray thing to be marked with FM_1 (figure 4.8c.) On the third pass through the loop, since gray thing bears FM_1 it cannot be colored with TM_1; this is due to the **off-head** condition on each link command that makes the markers mutually exclusive. No new nodes are marked in this pass, so the loop terminates. The last line of the algorithm marks any remaining unmarked nodes with QM_1 (figure 4.8d.)

As this example demonstrates, the upscan algorithm's T, F, and Q markers are in competition with each other. The T marker tries to propagate all the way up the IS-A hierarchy. Wherever the transitivity of a chain of IS-A links is to be broken by an exception, an F or Q marker on a node stops the T marker from getting in. Conversely, once a node has a T marker on it, it cannot be given an F or Q marker. T, F, and Q markers must be mutually exclusive in order to allow exceptions to override the transitivity of IS-A links.

111

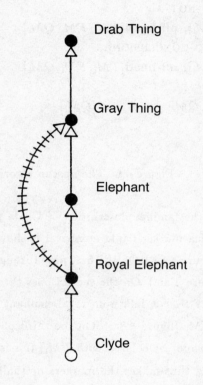

Figure 4.7: A description of Clyde the elephant.

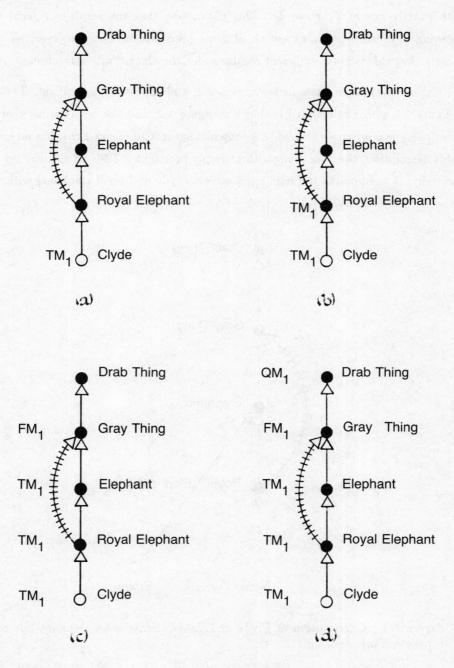

Figure 4.8: Computing the upscan of Clyde in four steps.

At the conclusion of the upscan procedure every node in the graph will be colored with exactly one of T, F, or Q. This guarantees that the result is a legal coloring, meaning it represents the extension of some predicate in $\mathcal{L}_\Pi$. However, we shall see shortly that this is not a sufficient condition for the algorithm's correctness.

The second upscan example features Ernie, a plain, ordinary elephant. The network of figure 4.7 with Ernie added is shown in figure 4.9, and the result of an upscan from Ernie using marker triple number 2 is shown in figure 4.10. Since Ernie has no exceptions in his description, the upscan algorithm simply propagates TM_2 all the way up the IS-A hierarchy. It then marks the remaining nodes (Clyde and royal elephant) with QM_2.

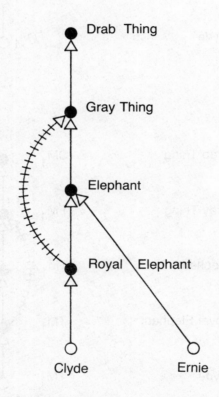

Figure 4.9: A description of Clyde and Ernie. Ernie is an elephant but not a royal elephant.

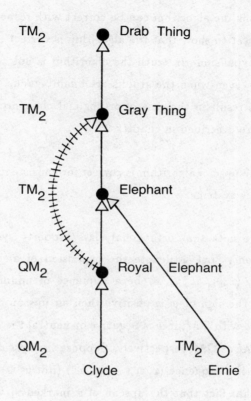

Figure 4.10: The result of an upscan of Ernie.

4.8 Correctness of the upscan algorithm

An upscan algorithm is correct with respect to a particular grounded expansion Φ iff the upscan of every node x produces the extension of A_x as determined by Φ. Since the upscan algorithm cannot distinguish between expansions, if a graph has multiple grounded expansions the algorithm can be correct with respect to at most one of them. However, we have yet to show that the algorithm is correct even when the graph has a unique grounded expansion. In truth the algorithm is not correct for general multiple inheritance graphs, even when the graphs are unambiguous. We can, however, prove a weaker correctness result in the case of orthogonal class/property multiple inheritance systems, which were described in chapter 2.

Theorem 4.1 The upscan algorithm is correct for consistent orthogonal class/property multiple inheritance systems.

Proof   First, we note that orthogonal class/property systems always have unique grounded expansions. Let Φ denote the expansion of an orthogonal class/property system Γ, and let $\langle x, y_1, \ldots, y_n, w \rangle$ be a sequence of minimal length in Φ such that $\langle x, w \rangle \in C(\Phi)$. If the sign of w is positive then an upscan of x should mark the node corresponding to w with TM; if w is negative or neutral the corresponding node should be marked with FM or QM, respectively. Suppose the upscan algorithm colors w incorrectly. Then there is a sequence $\langle x, z_1, \ldots, z_m, w' \rangle$ (not in Φ) representing this coloring, *i.e.* representing that fact that the upscan of x marked z_1 with TM, propagated TM from z_1 to z_2, and so on up to z_m, from which w was miscolored as w'. Since Γ is orthogonal, we have $y_i = z_i$ for $1 = i = min(n, m)$. Since Γ is consistent, $n \neq m$. If $n < m$ then the link $\langle z_n, w \rangle$ would have caused w to be colored correctly before z_m was reached and $\langle z_m, w' \rangle$ caused w to be mis-colored. Therefore, $m < n$. But that means z_m is an intermediary to $\langle x, y_1, \ldots, y_n, w \rangle$ in Φ, and Γ contains $\langle z_m, w' \rangle$, so $\langle x, y_1, \ldots, y_n, w \rangle$ is precluded and cannot appear in Φ.   ∎

Orthogonal class/property systems aren't the only ones for which the upscan algorithm works correctly. The reader should find it easy to prove that exception-free inheritance systems, which also have unique grounded expansions, work correctly with this algorithm.

116

To see why the upscan algorithm cannot be correct in the general case, consider its rule for implementing exceptions: the first of each triple of markers to reach a node prevents the other markers from entering. This is the shortest-path rule — a rule we rejected back in chapter 1 because it gave unreasonable results. Instead, we chose the inferential distance ordering. Since we are using a shortest path algorithm to reconstruct the conclusions of an expansion derived by inferential distance, we should expect the algorithm to sometimes produce incorrect results.

Figure 4.11 shows how it is possible for the wrong marker to win the race for a node. An upscan of Clyde in this network is shown in figures 4.12a through 4.12d. The upscan erroneously describes Clyde as gray thing and drab thing, due to marker TM_3 having reached gray thing from elephant before royal elephant could mark gray thing with FM_3.

Given a consistent network with a unique grounded expansion, an upscan will err only if the network contains a marker propagation path leading to an incorrect result that is shorter than all paths leading to the correct result. Comparing figure 4.11 with figure 4.7 shows that even a simple change in network topology can cause the upscan algorithm to err. If there are no redundant IS-A links, the upscan will never err since there is only one path along which a marker can propagate to any destination. (But the downscan algorithm for reconstructing E-predicates, discussed later, can produce invalid results even when there are no redundant links and no exceptions.) When redundant links are present, the upscan algorithm may work correctly some of the time; relative path lengths in the network are the determining factor.

Note that we cannot make the upscan algorithm work for general inheritance networks simply by removing redundant links, because as demonstrated in chapter 1, a link that is redundant with respect to one node may in fact not be redundant for a lower level node in the IS-A hierarchy due to the effects of exceptions.

4.9   PMPM's are still useful

The upscan algorithm we have chosen is incorrect in the general case of unrestricted inheritance networks. Given the limited computational abilities of a PMPM, it appears no PMPM algorithm could possibly compute the extensions of A-predicates correctly in unrestricted networks, and we will see later that the same holds for E-predicates.

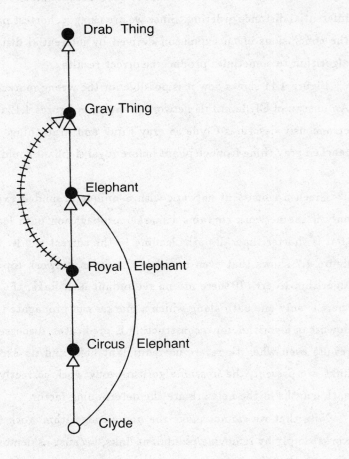

Figure 4.11: A network where an upscan of Clyde produces erroneous results.

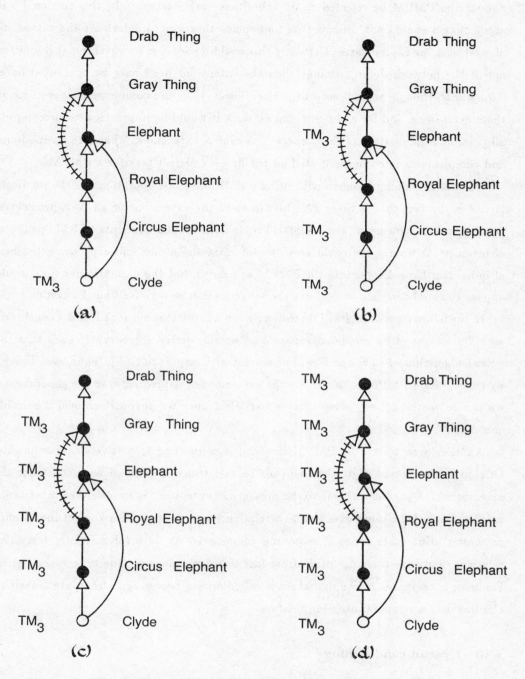

Figure 4.12: Computing the upscan of Clyde in four steps.

Should the PMPM be rejected as an inheritance architecture? In this section I will argue that it should not. Suppose we find some other way to determine the extensions of a set of A- or E-predicates. Certainly this could be done on a conventional computer, and if the network doesn't change then the extensions need only be computed once. Afterwards, though, we will need to perform meet, join, and complement operations on these extensions, and for any nontrivial network it would be impractical to precompute all possible combinations of these operations over N extensions. Therefore meets, joins, and complements must be computed on the fly — a perfect task for a PMPM.

The one remaining problem with using a PMPM for fast operations on the predicate lattice is the fact that it takes $3N^2$ bits to store the extensions of all N E-predicates, which is a lot of memory. For a 100,000 node network this amounts to 3.75 gigabtyes of memory. Of course, we could keep the information on disk and copy the extensions of individual E-predicates into the PMPM as needed, but the data transfer time would appear to equal the time to perform the set operation on a conventional machine.

In the following section I will show how, given a consistent network Γ and a conclusion set $C(\Phi)$ of one of its grounded expansions, we can derive a network Γ' such that the upscan algorithm when run on Γ' will be correct with respect to $C(\Phi)$. Thus, even though we cannot use a PMPM to determine the extensions of A-predicates in the general case, we can *reconstruct* those extensions on a PMPM once we have determined a suitable encoding Γ'.

Another area where the PMPM is useful is computing transitive closures quickly. Of course, the upscan algorithm depends on fast transitive closure to reconstruct the extensions of A-predicates. Due to the presence of exceptions in unrestricted inheritance graphs, transitive closure over the IS-A relation is no substitute for correct inheritance reasoning. But other types of reasoning common to AI do involve strictly transitive relations, and may therefore profit from fast transitive closure. One type suggested by Fahlman is temporal reasoning, where it is sometimes necessary to compute transitive closures over a temporal ordering relation.

4.10   Upscan conditioning

Given a consistent grounded expansion Φ, we desire a conditioned network Γ' such that the upscan algorithm run on Γ' will correctly reconstruct the extension of **every**

120

A-predicate in the set A_Π.

Theorem 4.2 If Γ is consistent then it is conditionable.

Proof First, set Γ' equal to $C(\Phi)$. Then, for every pair of nodes x and y such that x has no link to y, add $\langle +x, \#y \rangle$ to Γ'. The resulting network clearly satisfies our requirement that the upscan run correctly for all nodes, since an upscan of any node colors the entire graph on the first pass through the loop, and does so in strict accordance with $C(\Phi)$. ∎

The problem with the above method of constructing Γ' is that it generates a network with $O(N^2)$ links. Earlier we rejected as too expensive the idea of using $3N^2$ bits of memory to store the extensions of A-predicates explicitly. In that case, we certainly cannot afford to do the job with $O(N^2)$ links. We must find a more efficient encoding of $C(\Phi)$. Γ is presumably an efficient encoding; recall that one of the advantages of inheritance systems discussed in chapter 1 is that they provide for economy of representation; but Γ is not usable by the upscan algorithm. A reasonable approach then, might be to start with Γ and modify it as necessary, but no more than is necessary, to obtain a suitable Γ'. We shall call this modification *conditioning*.

An *additive* conditioning algorithm is one where $\Gamma' \supseteq \Gamma$. Non-additive conditioners may delete links from Γ as well as add them. In this thesis, for simplicity's sake, we will restrict the discussion to additive conditioners. Figure 4.13 is an additively conditioned version of figure 4.11. The upscan algorithm works correctly for all nodes in figure 4.13. The difference between this figure and the unconditioned network in figure 4.11 is one extra link.

The following algorithm for conditioning an IS-A acyclic network tries to add a minimal number of links. The algorithm checks each node, one at a time, to ensure that an upscan of that node produces the correct result. When a discrepancy in an upscan is discovered, it is corrected by adding another link. This is not an optimal conditioning algorithm, but it is efficient enough to be of practical use.

1. Determine the conclusion set $C(\Phi)$.

2. Set Γ' equal to Γ.

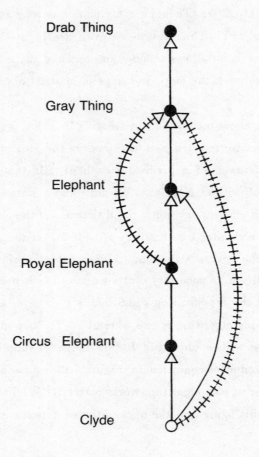

Figure 4.13: An additively conditioned version of figure 4.11

3. Pick a maximal node x (with respect to the $\prec$ ordering) that has not been picked before.

4. Run the upscan algorithm on node x in Γ'.

5. Find a closest node y (according to path length in Γ', not inferential distance) whose coloring does not agree with A_x. If no such node exists, return to step 3.

6. Put a link of the appropriate type from x to y into Γ' to force the upscan algorithm to color y correctly, and return to step 4.

The algorithm processes nodes in decreasing $\prec$ order, *i.e.* from highest to lowest in the IS-A hierarchy, in an attempt to minimize the number of links added. By time it is ready to condition a node x, all the conditioning links that will ever be added above x are already in place, so x has had all the help it is going to get from other nodes. At this point, if the upscan of x is still incorrect, only a direct link from x to the mis-colored node can fix it.

If we ran the conditioner on the network in figure 4.11, it would find that the upscans of drab thing, gray thing, elephant, royal elephant, and circus elephant all agreed with $C(\Phi)$, but the upscan of Clyde marks gray thing and drab thing with TM when gray thing should be marked with FM and drab thing with QM. Since gray thing is the closest mis-colored node in the upscan of Clyde, the algorithm adds an IS-NOT-A link from Clyde to gray thing to force gray thing to be colored with FM. When the algorithm is rerun on the modified network, the results are completely correct. No link need be added from Clyde to drab thing. The conditioned network is shown in figure 4.13.

This conditioning algorithm is efficient but not optimal because it considers nodes one at a time and only adds links where needed. Sometimes it is advantageous to add a link where it is not needed, because this saves adding several links at a lower level. For example, the network in figure 4.14a is a copy of figure 4.11, except that two more elephants, Ernie and Bertha, have been added. The network is not upscan conditioned. If the conditioner described above were run on this network, it would add three IS-NOT-A links to gray thing (one from each individual elephant) to make all the upscans work correctly. An optimal additive conditioner could accomplish the same thing by adding just one link, from circus elephant to gray thing, as shown in figure 4.14b. Even though

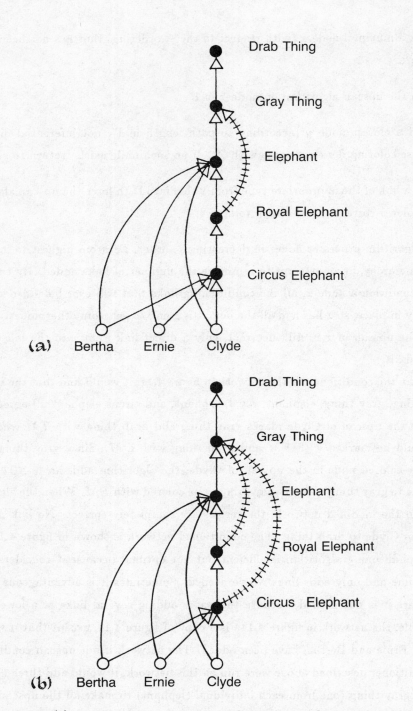

Figure 4.14: (a) an unconditioned version of figure 4.11 with two extra elephants; (b) an optimally conditioned version.

the upscan of circus elephant works correctly without this link, adding an IS-NOT-A link from circus elephant to gray thing saves three IS-NOT-A links from Clyde, Ernie, and Bertha to gray thing.

4.11 Effective conditioning

Many other conditioning algorithms are possible. I have written a MacLisp program called TINA (for Topological Inheritance Architecture) that finds the conclusion set of the grounded expansion of any consistent, unambiguous, i.a. network, and then conditions the network so that the upscan algorithm and its dual, the downscan, work correctly. TINA can also detect and report inconsistencies and ambiguities in the network. While TINA takes an additive approach to conditioning like the algorithm presented above, it does not run any simulated upscans or downscans. Instead it constructs lists of possible marker propagation paths; then it adds links to the network to block any paths that could cause incorrect results on a PMPM. Below is an example of TINA running on the network of figure 4.11:

```
*(load-net 'clyde1)                         ;Load the network shown
(C410DT50 CLYDE1 TIN)                       ;in figure 4.11.


*lnet                                       ;List the assertions.
((LINK-7 : CLYDE IS-A ELEPHANT)
 (LINK-6 : CLYDE IS-A CIRCUS-ELEPHANT)
 (LINK-5 : CIRCUS-ELEPHANT IS-A ROYAL-ELEPHANT
 (LINK-4 : ROYAL-ELEPHANT IS-A ELEPHANT)
 (LINK-3 : ROYAL-ELEPHANT IS-NOT-A GRAY-THING)
 (LINK-2 : ELEPHANT IS-A GRAY-THING)
 (LINK-1 : GRAY-THING IS-A DRAB-THING))


*(condition)                                ;Invoke TINA.
Computing the expansion.
Conditioning the upscan.
Added:  (CLYDE IS-NOT-A GRAY-THING) to fix upscan.
```

```
Conditioning the downscan.
Conditioning complete.
((CLYDE IS-NOT-A GRAY-THING))                    ;Value returned is list
                                                 ;of links TINA added.


*lnet                                            ;The revised network.
((LINK-8 : CLYDE IS-NOT-A GRAY-THING)
 (LINK-7 : CLYDE IS-A ELEPHANT)
 (LINK-6 : CLYDE IS-A CIRCUS-ELEPHANT)
 (LINK-5 : CIRCUS-ELEPHANT IS-A ROYAL-ELEPHANT)
 (LINK-4 : ROYAL-ELEPHANT IS-A ELEPHANT)
 (LINK-3 : ROYAL-ELEPHANT IS-NOT-A GRAY-THING)
 (LINK-2 : ELEPHANT IS-A GRAY-THING)
 (LINK-1 : GRAY-THING IS-A DRAB-THING))


*(plist 'clyde)                                  ;One of TINA's internal
(RELDOWNS (((CLYDE +)))                           ;data structures.

        IS-A-INHERITORS
        ((CLYDE))
        MKRPATHS
        (((GRAY-THING -))
         ((CIRCUS-ELEPHANT +))
         ((ELEPHANT +))
         ((CIRCUS-ELEPHANT +) (ROYAL-ELEPHANT +))
         ((CIRCUS-ELEPHANT +) (ROYAL-ELEPHANT +) (ELEPHANT +))
         ((CIRCUS-ELEPHANT +) (ROYAL-ELEPHANT +) (GRAY-THING -)))
        INHPATHS
        (((CIRCUS-ELEPHANT +))
         ((CIRCUS-ELEPHANT +) (ROYAL-ELEPHANT +))
         ((CIRCUS-ELEPHANT +)
          (ROYAL-ELEPHANT +)
```

126

```
      (GRAY-THING -))
     ((CIRCUS-ELEPHANT +) (ROYAL-ELEPHANT +) (ELEPHANT +)))
    IMMED
    ((GRAY-THING -) (CIRCUS-ELEPHANT +) (ELEPHANT +))
    SIGN
    NIL
    TOPSORT
    0)
```

An optimal conditioner would probably throw out Γ altogether and construct Γ' from scratch as a smallest correct encoding of $C(\Phi)$. I am not concerned with optimal conditioning algorithms here; I simply want to show that conditioning is feasible and not prohibitively expensive. The algorithm given in this chapter, which conditions a network by running simulated upscans on it, runs in time polynomial in the number of nodes. The number of links added is $O(N^2)$ in the worst case, but this is true even for optimal conditioners.

4.12   The downscan algorithm

A-predicates are the duals of E-predicates, as explained in chapter 3. They obey the symmetry relation $E_y(x) = A_x(y)$ for all $x, y \in \Pi$. However, the PMPM algorithm for reconstructing E-predicates, known as the *downscan* algorithm because it propagates markers downward along the IS-A hierarchy, is more complicated than the algorithm for reconstructing A-predicates. The downscan algorithm is more complicated because TM and FM must both propagate down the IS-A hierarchy. For an upscan, only TM must propagate upward. Consider figure 4.15. In 4.15a, the network indicates that A is not a B. It says nothing about whether A is a C or not; the fact that B is a C is irrelevant. An upscan of A would mark B with FM and leave C unmarked until the loop terminated and C was assigned QM by default. Contrast this with figure 4.15b. Here, A is a B, and B is not a C, so A is not a C. A downscan of C would mark B with FM due to the presence of the IS-NOT-A link, but it must mark A with FM too. Obviously, FM should propagate down the IS-A hierarchy the same as TM does.

Returning to figure 4.15a, a downscan of C in this network should mark B with

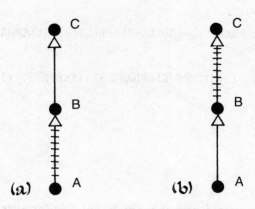

Figure 4.15: Two networks that illustrate the difference between upscan and downscan algorithms.

TM; it should then mark A with QM, not with FM. The heuristic we will adopt is: a downscan of node x puts FM on the tail node of any direct IS-NOT-A link to x. The FM's will be propagated downward through the IS-A hierarchy. For all other IS-NOT-A or NO-CONCLUSION links encountered during the scan, the tail node is colored with QM if the head node receives either TM or FM. The complete downscan algorithm is shown in figure 4.16.

Let's try out this algorithm on the graph in figure 4.7; the result is shown in figure 4.17. To perform a downscan of gray thing, we begin by placing TM_1 on gray thing and then FM_1 on royal-elephant. We then enter the loop. On the first iteration, TM_1 propagates downward from gray thing to elephant, and FM_1 propagates downward from royal-elephant to Clyde. On the second iteration, none of the commands inside the loop succeed, so the loop is exited. Finally, since drab thing has remained unmarked, it is marked with QM_1. We see that the set of gray things includes elephants, but not royal elephants or Clyde.

Now let's compute the downscan of drab thing using marker triple number 2; the result is figure 4.18. We begin by marking drab with TM_2. Then we enter the loop. In the first iteration we mark gray thing with TM_2. In the second we mark elephant with TM_2 and royal-elephant with QM_2. In the third iteration there are no nodes left to mark, since QM_2 doesn't propagate across IS-A links. Clyde has remained unmarked, so after the loop terminates he is marked with QM_2. We see that the set of drab things includes

```
procedure downscan(x: node; i: marker-triple) = begin
  clear[TM_i, FM_i, QM_i];
  name[x]  ⟹  set[TM_i];
  link-type[IS-NOT-A], on-head[TM_i]  ⟹  set-tail[FM_i];
  loop
    link-type[IS-NOT-A, NO-CONCLUSION],
      any-on-head[TM_i, FM_i], off-tail[TM_i, FM_i, QM_i]  ⟹  set-tail[QM_i];
    link-type[IS-A],
      on-head[TM_i], off-tail[TM_i, FM_i, QM_i]  ⟹  set-tail[TM_i];
    link-type[IS-A],
      on-head[FM_i], off-tail[TM_i, FM_i, QM_i]  ⟹  set-tail[FM_i]
  endloop;
  off[TM_i, FM_i, QM_i]  ⟹  set[QM_i]
end
```

Figure 4.16: The downscan algorithm.

gray things and elephants, but there is no conclusion about whether royal elephants or Clyde in particular are drab.

These examples have all been too small to demonstrate the real power of massively parallel computation. We could add a few more elephants to the inheritance graph, or a few thousand more — as long as depth remained the same as that of figure 4.5 the downscan algorithm would take *no* additional time to reconstruct the extension of any predicate in E_Π.

4.13 Correctness of the downscan algorithm

Like the upscan algorithm, the downscan algorithm is not correct for general multiple inheritance graphs. Unlike upscan, it is not correct even when the network has no redundant links and no exceptions. We can, however, prove correctness for some restricted cases.

Theorem 4.3 The downscan algorithm is correct for consistent orthogonal class/property inheritance systems.

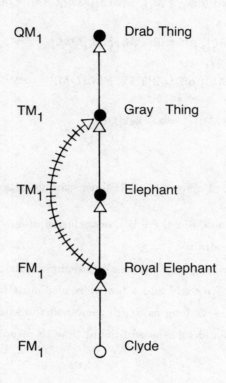

Figure 4.17: Result of a downscan of gray thing.

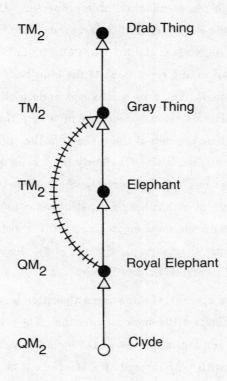

Figure 4.18: Result of a downscan of drab thing.

Proof   We divide downscans into three cases, according to the type of node whose E-predicate is being reconstructed. Individual nodes are trivial: the downscan simply marks the individual with TM and everything else with QM. Predicate nodes of type C, representing classes, are an only slightly less trivial case, since such nodes have nothing but IS-A links below them; the downscan in this case is a simple transitive closure. Clearly the algorithm is correct for these two cases. The remaining case covers predicate nodes of type P, which represent inheritable properties. Consider the downscan of a predicate node w of type P. The algorithm first sends TM, FM, and QM marks from w downward across IS-A, IS-NOT-A, and NO-CONCLUSION links, respectively. All the nodes marked in this step and in any repetitions of the loop body will be either predicates of type C, or else individuals. Let x be a maximal node with respect to the $\prec$ ordering, and let $\langle x, y_1, \ldots, y_n, w \rangle$ be a shortest sequence in Φ such that the downscan of w colors x incorrectly. (Recall that the sign of the token w in the above sequence determines the color that x should be marked with.) Obviously there is no direct link from x to w, since if there were x could not be colored incorrectly. Since the network is orthogonal, y_1 is the only immediate superior of x that lies on a path from x to w. Therefore the downscan must have colored y_1 with the same marker as x. But Φ also contains $\langle y_1, \ldots, y_n, w \rangle$, so if x is colored incorrectly then so is y_1. Since $x \prec y_1$, this contradicts our assumption that x was a maximal incorrectly colored node.   ∎

Unlike in the case of upscans, the downscan algorithm is *not* correct for exception-free inheritance systems. Figure 4.19a demonstrates this. The graph has a unique, consistent expansion in which A is a D, *i.e.* the set of D's includes A. Yet a downscan of D in this graph would mark A with QM instead of TM. Note that the graph is upscan conditioned: the upscan algorithm produces the correct results for every node. Obviously, then, upscan conditioning does not assure downscan conditioning. Figure 4.19b shows a downscan conditioned network derived from 4.19a. The algorithm for downscan conditioning is, *mutatis mutandis*, the same as for upscan conditioning, except that nodes should be processed in increasing $\prec$ order (*i.e.* from lowest to highest with respect the IS-A hierarchy.)

In order to accurately reconstruct both A-predicates and E-predicates, the network must be both upscan and downscan conditioned. But the links that one conditioner adds can upset the complementary scan, thus producing more work for the complementary

132

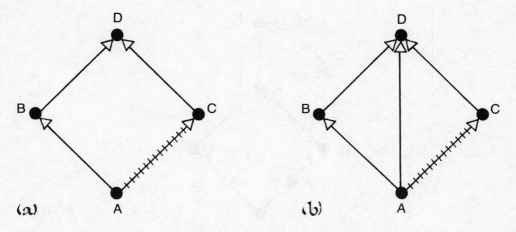

Figure 4.19: (a) a consistent, unambiguous, upscan conditioned network that is not downscan conditioned; (b) a version that is both upscan and downscan conditioned.

conditioner. Figure 4.20a is an example. When we run the upscan conditioner on this network, it finds that the network is properly conditioned and needs no additional links. Then, the downscan conditioner discovers that an IS-A link must be put in between A and E to make the downscan of E work correctly. The result is figure 4.20b. Since the downscan conditioner has modified the network, we must run the upscan conditioner on the new version. The upscan conditioner finds that the network is no longer upscan conditioned. It adds an IS-NOT-A link from A to F to fix the upscan of A. The result is figure 4.20c. Finally the downscan conditioner is run again; it finds no need for further changes.

The interaction between upscan and downscan conditioning might lead one to wonder whether the process always terminates. With non-additive conditioners an infinite loop may be possible, depending on the algorithm used, but since additive conditioners move monotonically toward a complete inheritance graph (one with $N^2 - N$ links), additive conditioning is guaranteed to terminate.

An example of TINA doing downscan conditioning on the network of figure 4.20a is shown below:

```
*(load-net 'ch4b)                        ;Load network shown
(C410DT50 CH4B TIN)                      ;in figure 4.20a.
```

133

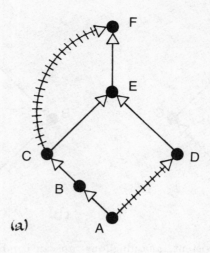

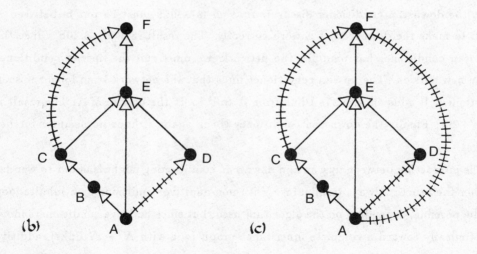

Figure 4.20: Links that one conditioner adds can upset the complementary scan, as in this network; after (a) upscan conditioning then (b) downscan conditioning it requires (c) upscan conditioning again.

```
*lnet                                    ;List the assertions.
((LINK-7 : E IS-A F)
 (LINK-6 : D IS-A E)
 (LINK-5 : C IS-A E)
 (LINK-4 : C IS-NOT-A F)
 (LINK-3 : B IS-A C)
 (LINK-2 : A IS-NOT-A D)
 (LINK-1 : A IS-A B))


*(condition)                             ;Invoke TINA.
Computing the expansion.
Conditioning the upscan.
Conditioning the downscan.
Added:  (A IS-A E) to fix downscan.
Conditioning the upscan.
Added:  (A IS-NOT-A F) to fix upscan.
Conditioning the downscan.
Conditioning complete.                   ;Value returned is list
((A IS-NOT-A F) (A IS-A E))              ;of links TINA added.


*lnet                                    ;The revised network.
((LINK-9 : A IS-NOT-A F)
 (LINK-8 : A IS-A E)
 (LINK-7 : E IS-A F)
 (LINK-6 : D IS-A E)
 (LINK-5 : C IS-A E)
 (LINK-4 : C IS-NOT-A F)
 (LINK-3 : B IS-A C)
 (LINK-2 : A IS-NOT-A D)
 (LINK-1 : A IS-A B))
```

135

4.14 Relationship between Γ and Γ′

Let Γ′ be an additively conditioned version of Γ. What can be said about the grounded expansions of Γ′? Figure 4.21 shows that Γ′ may have fewer grounded expansions than Γ. The network in figure 4.21a has two grounded expansions. Suppose we choose the one whose conclusion set contains $\langle +p, -q \rangle$. Then we will need to add a direct link between P and Q to make the upscan come out right. The conditioned network is shown in figure 4.21b: this network has only one grounded expansion.

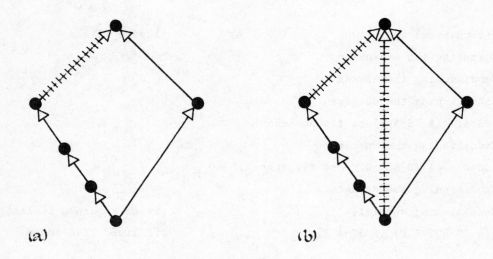

(a)   (b)

Figure 4.21: (a) a network with two grounded expansions; (b) a conditioned version with only one grounded expansion.

Could additive conditioning lead to an *increase* in the number of grounded expansions? I have no answer at the present time, but I suspect not. Conjecture: every grounded expansion $\Phi′$ of an additively conditioned Γ′ is a superset of some grounded expansion Φ of Γ. This conjecture would be false if we had settled on forward chained rather than doubly chained inheritance (defined in chapter 2.)

```
procedure mark-dubious(p, q: node; DU, DD : marker) = begin
  clear[DU, DD];
  name[p]   ⟹   set[DU, DD];
  name[q]   ⟹   set[DD];
  loop
    link-type[IS-A], on-head[DU], off-tail[DU]   ⟹   set-tail[DU];
    link-type[IS-A], on-tail[DD], off-head[DD]   ⟹   set-head[DD]
  endloop;
  link-type[IS-NOT-A], on-tail[DD]   ⟹   set-head[DD]
end
```

Figure 4.22: The procedure for finding nodes affected by an update.

4.15 Updating the knowledge base

Any change to the knowledge base is likely to result in a change to the expansion. However, even if the expansion remains the same, the change may damage the conditioning
of the network. Therefore, in order to insure the correctness of the upscan and downscan
algorithms, the network should be reconditioned after every update. Obviously this is
a disappointing result, since conditioning is quite expensive compared to the cost of a
single marker propagation scan.

We can soften the blow somewhat by observing that adding or deleting a link from
node P to node Q affects only a certain well-defined slice of the network. Only nodes
below P can have their upscan affected. Only nodes above P or Q can have their
downscan affected — and here we must take IS-NOT-A links into account too. Let us
use an individual marker DU (for Dubious Upscan) to mark every node whose upscan
might be affected by adding or deleting a link between nodes P and Q. Similarly, we
will use an individual marker DD (for Dubious Downscan) to mark every node whose
downscan might be affected. Figure 4.22 shows the procedure for finding the affected
nodes.

We cannot perform an upscan of any node with DU or a downscan of any node
with DD until the network has been reconditioned, but in the interim we remain free to
perform upscans and downscans on other nodes. While a serial machine performs the

computation for reconditioning the affected slice of network, the PMPM can continue to service requests for upscans and downscans of nodes in unrelated parts of the network. (Though one might ask how often one adds information and then doesn't make use of it soon afterwards.)

Another solution might be to delay the actual addition or deletion of a link until the results of the reconditioning are ready. Then the worst that can happen if we perform an upscan or downscan of a node marked with *DU* or *DD*, respectively, is that the results will reflect the state of the knowledge base prior to the update. The function of the **mark-dubious** scan in this case is to flag any nodes whose PMPM description is lagging behind the current state of knowledge. The program making requests of the PMPM can decide whether this lag is important to a given query.

5 A Theory of Inheritable Relations

Q: What's big and gray, has a trunk, and lives in the trees?
A: An elephant. I lied about the trees.

— Ron Brachman

5.1 Introduction

In this chapter I introduce notation for a more advanced topic in inheritance reasoning: inheritable relations with exceptions. These are binary relations that inherit along the IS-A hierarchy (Touretzky, 1985). After presenting some examples, I define sequential and graphical notations for such relations, formalize the underlying inference rule, and analyze the result as an extension to the generic inheritance system of chapter 2. The structure of this chapter parallels that of chapter 2. Many of the definitions are repeated verbatim, but now they have expanded meanings. Two factors that complicate the treatment of relations, just as they complicated the treatment of IS-A inheritance, are the need to provide for exceptions and the need to control interactions between competing inferences due to multiple inheritance and, now, multiple relational statements.

5.2 An example of an inheritable relation

In the sentence "Clyde loves Fred," loves is an ordinary relation that holds between two individuals. In the sentence "elephants love zookeepers," loves is an inheritable relation. From this sentence plus the assertion that Clyde is an elephant we can infer that Clyde loves zookeepers. If we add that Fred is a zookeeper, we can infer that Clyde loves Fred. The argument looks like this:

Elephants love zookeepers.

Clyde is an elephant.

Therefore, Clyde loves zookeepers.

Fred is a zookeeper.

Therefore, Clyde loves Fred.

An alternative derivation of "Clyde loves Fred" from the same set of axioms is:

> Elephants love zookeepers.
>
> Fred is a zookeeper.
>
> Therefore, elephants love Fred.
>
> Clyde is an elephant.
>
> Therefore, Clyde loves Fred.

Relational inheritance is obviously dependent on IS-A hierarchy inheritance as studied in chapter 2, since the IS-A hierarchy determines the paths along which relations can inherit. But relational inheritance is more complex because it involves multiple slices of the hierarchy at once: one slice for each argument to the relation. To derive "Clyde loves Fred" from "elephants love zookeepers," for example, we must thread the relation along two inheritance paths. One path runs from Clyde to elephant, the other from Fred to zookeeper.

5.3 Relations in predicate logic

The language of inheritable relations defined in this thesis allows assertions only about individuals and atomic classes, not their complements, unions, or intersections. This does not preclude negative statements, however. "Elephants don't love zookeepers" is acceptable; "elephants love non-zookeepers" is not. Future extensions to this system might permit the latter statement.

A statement about two individuals, such as "Clyde loves Fred," can be expressed in the predicate calculus using a two-place predicate, *viz.* Loves(clyde, fred). Ignoring exceptions for the moment, "elephants love zookeepers" can be expressed as:

$$(\forall x, y) \quad \text{Elephant}(x) \wedge \text{Zookeeper}(y) \rightarrow \text{Loves}(x, y)$$

Consequently, "elephants love Fred" and "Clyde loves zookeepers" are expressible as:

$$(\forall x) \quad \text{Elephant}(x) \rightarrow \text{Loves}(x, \text{fred})$$

$$(\forall y) \quad \text{Zookeeper}(y) \rightarrow \text{Loves}(\text{clyde}, y)$$

140

5.4 Frames, slots, and relations

In chapter 1 we saw that when frames are translated into logical language, slots become partial functions. Thus, to say that Clyde is gray, we would write $color(clyde) = gray$, and to say that elephants are gray (without exception) we would write

$$(\forall x) \quad \text{Elephant}(x) \rightarrow \text{color}(x) = \text{gray}.$$

In chapter 2 we adopted a simplified view of frames in which slot values were expressed as predicates. "Clyde is gray" became GrayThing(clyde) and "elephants are gray" became

$$(\forall x) \quad \text{Elephant}(x) \rightarrow \text{GrayThing}(x).$$

The simplification allowed us to express "elephants are gray" the same way as "elephants are mammals." In this chapter we will examine an alternative formalization of slots: as inheritable relations. We will now express "Clyde is gray" by Color(clyde, gray), and "elephants are gray" by

$$(\forall x) \quad \text{Elephant}(x) \rightarrow \text{Color}(x, \text{gray}).$$

5.5 Exceptions to relations

Relational statements are subject to exceptions just as IS-A hierarchy statements are. Consider the following:

Citizens dislike crooks.
Gullible citizens are citizens.
Elected crooks are crooks.
Gullible citizens do not dislike elected crooks.

An inheritable relation (dislikes) can be overridden by an exception in the form of a contradictory relation (does not dislike). We might call this type of exception a relational exception. The inheritance of relations is also governed by the more familiar IS-A hierarchy exception explored in chapter 2, since inheritable relations are dependent upon the IS-A hierarchy. For example:

Citizens dislike crooks.

Politicians are crooks.

Nixon is a politician.

But Nixon is not a crook.

No inference about whether citizens dislike Nixon.

I hope the reader has seen enough examples by now to desire a more formal treatment of inheritable relations. We shall develop one in the remainder of this chapter. As was the case in chapter 2, our task is complicated by the issues of exceptions and multiple inheritance. In the following sections we will define representations for relational assertions in both sequential and graphical form, and then develop an inferential distance ordering for relations.

5.6 Extending Π and Θ to include relations

The set of concepts, Π, defined in chapter 2 consisted of individual and predicate nodes. We now expand Π to also include a class of derived predicates known as *relational* predicates. A relational predicate is a unary predicate derived from an n-ary relation by supplying values for $n-1$ of the arguments. In this chapter we will restrict the discussion to binary relations, but the technique easily extends to relations of higher arity. We will use a variant of lambda notation for abbreviating the definitions of unary predicates derived from binary ones; the reason for the nonstandard notation will become clear shortly. Let us consider the loves relation, written Loves(x, y). If we were using ordinary first order logic we could construct two types of unary predicates from Loves(x, y) by substituting individuals for one or the other of the arguments. For example, given an individual a, we can abbreviate the two derived unary predicates shown below. The arguments are enclosed in brackets rather than parentheses as a reminder that LOVES[] isn't itself a predicate, but rather shorthand notation for a lambda expression that defines a predicate.

Abbreviation	Informal Interpretation	FOL Predicate
LOVES$[a]$	Things that love a.	$\lambda x.\,\text{Loves}(x, a)$
LOVES$^{-1}[a]$	Things loved by a.	$\lambda x.\,\text{Loves}(a, x)$

142

Since we are not working in ordinary first order logic but rather in a form of nonmonotonic logic that admits statements about classes as well as about individuals, we shall specify four types of relational predicates instead of just two. Let a be an individual and p a predicate. Here are the four types of relational LOVES predicates we will admit to our inheritance language, along with their informal interpretations. Again, we enclose the arguments in brackets rather than parentheses as a reminder that we are expressing a complex lambda expression (*i.e.*, complex when written in nonmonotonic logic) in a shorthand form. The nonmonotonic logic forms of these relational predicates will be given in the next section.

Abbreviation	Informal Interpretation
LOVES[+a]	Things that love a.
LOVES[+p]	Things that love p's.
LOVES^{-1}[+a]	Things loved by a.
LOVES^{-1}[+p]	Things loved by p's.

If we substitute negative or neutral predicate tokens instead of positive predicate tokens for arguments, we can obtain other types of relational predicates. Since their treatment would complicate the definitions and proofs that follow, it will be omitted from this thesis, except that we shall list the predicates here:

Abbreviation	Informal Interpretation
LOVES[−p]	Things that love non-p's.
LOVES^{-1}[−p]	Things loved by non-p's.
LOVES[#p]	Things that love things whose p-ness is inconclusive.
LOVES^{-1}[#p]	Things loved by things whose p-ness is inconclusive.

Recall that Θ, the set of tokens, was defined in chapter 2 as the cartesian product of $\{+, -, \#\}$ with the individuals and predicates of Π. Now that Π also contains relational predicates, Θ will contain relational tokens. Two examples are +LOVES[+fred], which denotes things that love Fred, and −LOVES[+fred], which denotes things that don't love Fred. To get slightly ahead of ourselves, I will reveal here that we express "Clyde loves Fred" in this notation by the sequence $\langle$+clyde, +LOVES[+fred]$\rangle$. In other words, Clyde is a member of the class "things that love Fred." We could express the same idea another way by writing $\langle +\text{fred}, +\text{LOVES}^{-1}[+\text{clyde}]\rangle$, *i.e.* Fred is a member of the class "things that Clyde loves." The two sequences above are mutual inverses. We will give a formal definition for the inverses of sequences shortly.

5.7 Relational tokens as nonmonotonic logic expressions

There are twelve types of relational tokens in the inheritance language. These are derived from the cartesian product of three signs $(+, -, \#)$ and the four types of relational predicates listed previously. We describe six of the twelve here by giving their formalization in Moore's autoepistemic version of nonmonotonic logic. The remaining six tokens are identical to the first six with their arguments reversed. Note that as in chapter 2, our formalization in nonmonotonic logic is incomplete because it does not take the inferential distance ordering into account.

Let R be a relation, b an individual, and q a predicate. The first six types of admissible relational tokens are:

Relational Token	Nonmonotonic Logic Interpretation
$+R[+b]$	$\lambda x.R(x,b)$
$-R[+b]$	$\lambda x. \sim R(x,b)$
$\#R[+b]$	$\lambda x.M[R(x,b)] \wedge M[\sim R(x,b)]$
$+R[+q]$	$\lambda x.(\forall y)q(y) \wedge M[R(x,y)] \rightarrow R(x,y)$
$-R[+q]$	$\lambda x.(\forall y)q(y) \wedge M[\sim R(x,y)] \rightarrow \sim R(x,y)$
$\#R[+q]$	$\lambda x.(\forall y)q(y) \wedge M[M[R(x,y)] \wedge M[\sim R(x,y)]] \rightarrow$ $M[R(x,y)] \wedge M[\sim R(x,y)]$

The formalization of the remaining six types of relational tokens is left as an exercise for the reader. As a second exercise, try formalizing relational tokens corresponding to the relational predicates we excluded from the inheritance language, *i.e.* those where negative or neutral rather than positive predicate tokens were substituted for arguments.

5.8 Relational sequences

Having extended Θ, the set of all tokens, by adding relational tokens, we can now extend the language of inheritance assertions (a subset of $\Theta \times \Theta$) by adding a new form of ordered pair, called a relational pair. There are twenty-four types of well-formed relational pairs.

These twenty-four are the product of two types of ordinary token in the first position of the ordered pair (either a positive individual token or a positive predicate token) and twelve types of relational token in the second position. Let R be a binary relation $R(x, y)$, let a and b be individuals, and let p and q be predicates. Here are the first twelve of the twenty-four types of well-formed ordered pairs involving relations, with their informal interpretations:

Sequence	Informal Interpretation
$\langle +a, +R[+b] \rangle$	a is R to b.
$\langle +a, -R[+b] \rangle$	a is not R to b.
$\langle +a, \#R[+b] \rangle$	No conclusion whether a is R to b.
$\langle +a, +R[+q] \rangle$	a is R to q's.
$\langle +a, -R[+q] \rangle$	a is not R to q's.
$\langle +a, \#R[+q] \rangle$	No conclusion whether a is R to q's.
$\langle +p, +R[+b] \rangle$	p's are R to b.
$\langle +p, -R[+b] \rangle$	p's are not R to b.
$\langle +p, \#R[+b] \rangle$	No conclusion whether p's are R to b.
$\langle +p, +R[+q] \rangle$	p's are R to q's.
$\langle +p, -R[+q] \rangle$	p's are not R to q's.
$\langle +p, \#R[+q] \rangle$	No conclusion whether p's are R to q's.

The remaining twelve types are identical to the above except in that they use the inverse form of the relation, causing the lambda variable to appear in the second argument position rather than the first.

5.9 Relational ordered pairs as logical sentences

In this section we give nonmonotonic logic interpretations for the first twelve of the twenty four types of admissible relational assertions. The reader is invited to compare these *sentences* in nonmonotonic logic with the six nonmonotonic logic *expressions* for relational predicates given earlier, from which the sentences are derived.

145

Sequence	Nonmonotonic Logic Interpretation
$\langle +a, +R[+b] \rangle$	$R(a, b)$
$\langle +a, -R[+b] \rangle$	$\sim R(a, b)$
$\langle +a, \#R[+b] \rangle$	$M[R(a, b)] \wedge M[\sim R(a, b)]$
$\langle +a, +R[+q] \rangle$	$(\forall y) \quad q(y) \wedge M[R(a, y)] \rightarrow R(a, y)$
$\langle +a, -R[+q] \rangle$	$(\forall y) \quad q(y) \wedge M[\sim R(a, y)] \rightarrow \sim R(a, y)$
$\langle +a, \#R[+q] \rangle$	$(\forall y) \quad q(y) \wedge M[M[R(a, y)] \wedge M[\sim R(a, y)]] \rightarrow$ $M[R(a, y)] \wedge M[\sim R(a, y)]$
$\langle +p, +R[+b] \rangle$	$(\forall x) \quad p(x) \wedge M[R(x, b)] \rightarrow R(x, b)$
$\langle +p, -R[+b] \rangle$	$(\forall x) \quad p(x) \wedge M[\sim R(x, b)] \rightarrow \sim R(x, b)$
$\langle +p, \#R[+q] \rangle$	$(\forall x) \quad p(x) \wedge M[M[R(x, b)] \wedge M[\sim R(x, b)]] \rightarrow$ $M[R(x, b)] \wedge M[\sim R(x, b)]$
$\langle +p, +R[+q] \rangle$	$(\forall x, y) \quad p(x) \wedge q(y) \wedge M[R(x, y)] \rightarrow R(x, y)$
$\langle +p, -R[+q] \rangle$	$(\forall x, y) \quad p(x) \wedge q(y) \wedge M[\sim R(x, y)] \rightarrow \sim R(x, y)$
$\langle +p, \#R[+q] \rangle$	$(\forall x, y) \quad p(x) \wedge q(y) \wedge M[M[R(x, y)] \wedge M[\sim R(x, y)]] \rightarrow$ $M[R(x, y)] \wedge M[\sim R(x, y)]$

5.10 Relational inheritance paths

As in chapter 2, sequences of length greater than two are not part of the inheritance
language itself but represent inheritance paths. For example, the relational sequence

$$\langle +\text{clyde}, +\text{elephant}, +\text{LOVES}[+\text{zookeeper}], +\text{LOVES}[+\text{fred}] \rangle$$

represents the inference that Clyde is an elephant and therefore a lover of zookeepers,
and therefore a lover of Fred.

5.11 Inverses of sequences

A relational sequence and its inverse are distinct sequences, but their meanings are
identical. One can think of a sequence σ and its inverse σ^{-1} as being active and passive
versions of the same sentence. We define an inverse sequence σ^{-1} for every *inheritable*
sequence $\sigma \in \Sigma$ as follows. The inverse of the sequence

$$\sigma = \langle x_1, \ldots, x_n, r[y_m], \ldots, r[y_1] \rangle$$

146

is defined to be

$$\sigma^{-1} = \langle y_1, \ldots, y_m, r^{-1}[x_n], \ldots, r^{-1}[x_1] \rangle$$

when $m > 0$ (*i.e.* the sequence contains at least one relational token.) Otherwise, when σ is of form $\langle x_1, \ldots, x_n \rangle$ (meaning it contains no relational tokens, so $m = 0$) then σ^{-1} is defined to be σ. We shall prove later that every sequence that can appear in a grounded expansion is of the form $\langle x_1, \ldots, x_n, r[y_m], \ldots, r[y_1] \rangle$ (with m possibly 0.)

Note that the inverses of well-formed relational assertions are themselves well-formed, while the inverses of disallowed relational assertions are also disallowed, *e.g.* the sequence $\langle +a, +R[-p] \rangle$, which is disallowed because the argument of the relation is a negative token, has as its inverse the sequence $\langle -p, +R[+a] \rangle$, which is disallowed because the first component of the sequence is a negative token.

Examples of inverses: $\langle +\text{clyde}, +\text{elephant}, +\text{mammal} \rangle$ is its own inverse. The inverse of the sequence

$$\langle +\text{elephant}, +\text{LOVES}[+\text{zookeeper}] \rangle,$$

meaning elephants are lovers of zookeepers, is

$$\langle +\text{zookeeper}, +\text{LOVES}^{-1}[\text{ elephant}] \rangle,$$

which means zookeepers are loved by elephants. The inverse of

$$\langle +\text{clyde}, +\text{elephant}, +\text{LOVES}[+\text{zookeeper}], +\text{LOVES}[+\text{fred}] \rangle$$

which means Clyde is an elephant, therefore a lover of zookeepers, and therefore a lover of Fred, is

$$\langle +\text{fred}, +\text{zookeeper}, +\text{LOVES}^{-1}[+\text{elephant}], +\text{LOVES}^{-1}[+\text{clyde}] \rangle$$

which means Fred is a zookeeper, and therefore is loved by elephants, and therefore is loved by Clyde.

5.12 The graphical representation of relations

For each binary relation R we define three link types, $+R$, $-R$, and $\#R$ to represent the twelve forms of relational statements and their inverses. We will draw relational links similarly to inheritance links, except that they will be drawn horizontally, with open

+ LOVES

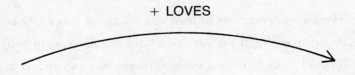

− LOVES

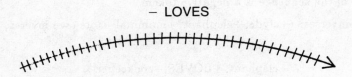

LOVES

Figure 5.1: The three link types for the LOVES relation.

arrowheads, and labelled with the signed relation name. Figure 5.1 shows the three link types for the loves relation. Figure 5.2 shows the network representation of "citizens dislike crooks, but gullible citizens do not dislike elected crooks." The figure also contains two individuals, Fred and Dick. In sequence notation the network is described by:

⟨+citizen, +DISLIKES[+crook]⟩

⟨+gullible.citizen, +citizen⟩

⟨+elected.crook, +crook⟩

⟨+gullible.citizen, −DISLIKES[+elected.crook]⟩

⟨+fred, +gullible.citizen⟩

⟨+dick, +elected.crook⟩

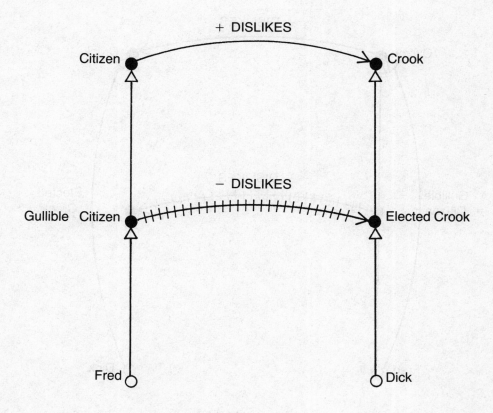

Figure 5.2: Citizens dislike crooks, but gullible citizens do not dislike elected crooks.

149

Fred is a gullible citizen and Dick is an elected crook. By inheritance, Fred dislikes crooks but Fred does not dislike Dick.

5.13 The inferential distance ordering applied to relations

In the example above, we infer by inheritance that Fred is a citizen and Dick a crook. If we make these facts explicit in the network, the result is figure 5.3. As before, citizens dislike crooks. Then why doesn't Fred dislike Dick? The answer has to do with the inferential distance ordering. Fred is a special type of citizen: a *gullible* citizen, and Dick is a special type of crook: an *elected* crook. The relation between citizens and crooks in general is overridden by a different one between gullible citizens and elected crooks.

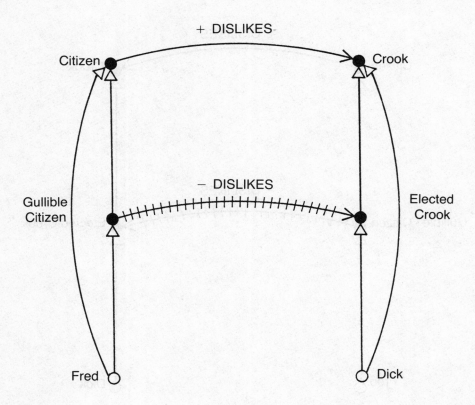

Figure 5.3: Adding redundant IS-A links shouldn't change the meaning of the network.

Inferential distance is only a *partial* ordering. Just as there were situations in IS-A hierarchy inheritance where ambiguity led to multiple expansions (as in figure 2.7 in chapter 2), there are situations in relational inheritance that are equally ambiguous. One example is figure 5.4, shown in sequence form below. Elephants are mammals and snakes are reptiles. Given that mammals typically fear snakes and elephants typically do not fear reptiles, how do elephants feel about snakes?

⟨+elephant, +mammal⟩
⟨+snake, +reptile⟩
⟨+mammal, +FEARS[+snake]⟩
⟨+elephant, −FEARS[+reptile]⟩

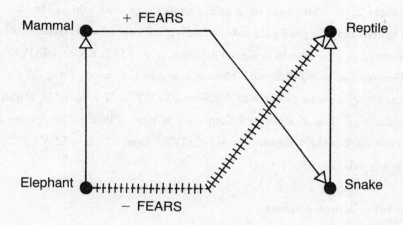

Figure 5.4: A network with two consistent grounded expansions. Do elephants fear snakes, or not?

One line of argument, corresponding to one of the expansions, says that since elephants do not fear reptiles, and snakes are reptiles, elephants must not fear snakes. On the other hand (rather, in the other expansion), since elephants are mammals and mammals fear snakes, elephants must fear snakes. These two expansions are mutually contradictory and there is no basis for choosing one over the other.

Note that in figure 5.2, where the link from citizen to crook is overridden by one from gullible citizen to elected crook, gullible citizen is below citizen, and elected crook is below crook. In figure 5.4, where the link from elephant to reptile fails to override or be overridden by the link from mammal to snake, elephant is below mammal but reptile is above snake. In general, for the inferential distance ordering to apply, the head and tail of the overriding relational link must be at or below the head and tail nodes, respectively, of the link being overridden. We will formalize this rule shortly.

5.14   Notation

To review, Π denotes the set of all individuals, real-world predicates, and derived relational predicates. Θ is the set of all individual, predicate, and relational tokens. Σ is the set of all sequences over Θ of length at least two, and σ denotes an element of Σ. Γ denotes a set of inheritance assertions, *i.e.* $\Gamma \subseteq \Theta \times \Theta \subseteq \Sigma$. Φ and S also denote subsets of Σ.

As in chapter 2, variables such as x and y shall range over tokens. In this chapter, though, the tokens may be predicate tokens, individual tokens, or relational tokens. We will use the variable r to denote a signed relation, *e.g.* +LOVES or +LOVES^{-1}. The prime notation applies to relation variables such as r in the obvious way: if r stands for +LOVES, then r' can mean either −LOVES or #LOVES. If a variable x stands for a relational token $r[y]$, then x' shall be interpreted as $r'[y]$. Finally, the inverse notation applied to relation variables means if r is +LOVES then r^{-1} is +LOVES^{-1} and vice versa. If r is any relation, $(r^{-1})^{-1}$ is r.

5.15   The inheritance axioms

The definitions in this section are for the most part taken verbatim from chapter 2. However, their meanings have been expanded by the addition of relational tokens. Most definitions are followed by an example of how they apply to relational tokens.

Definition 5.1 The *conclusion set* $C(\Phi)$ of a set of sequences Φ is defined as $C(\Phi) = \{\langle x, y \rangle \mid \langle x, \ldots, y \rangle \in \Phi\}$.

Definition 5.2 A set of sequences Φ *contradicts* the sequence $\langle x_1, \ldots, x_n \rangle$ iff $\langle x_1, x_i' \rangle \in C(\Phi)$ for some i, $1 \le i \le n$.

Example: if Φ contains the sequence

$$\langle +\text{elephant}, +\text{mammal}, +\text{FEARS}[+\text{snake}]\rangle,$$

which means elephants are mammals and therefore fear snakes, then it contradicts the sequence

$$\langle +\text{elephant}, -\text{FEARS}[+\text{reptile}], -\text{FEARS}[+\text{snake}]\rangle,$$

which means elephants do not fear reptiles and therefore do not fear snakes.

Definition 5.3 A token z is an *intermediary* in Φ to a sequence

$$\langle x_1, \ldots, x_n, r[y_m], \ldots, r[y_1]\rangle$$

iff either $z = x_i$ for some i between 1 and n, or Φ contains a sequence

$$\langle x_1, \ldots, x_i, z_1, \ldots, z_p, x_{i+1}\rangle$$

where z is one of $z_1 \ldots z_p$ and $1 \le i < n$.

Example: If Φ contains the sequence

$$\langle +\text{fred}, +\text{gullible.citizen}, +\text{citizen}\rangle$$

then $+$gullible.citizen is an intermediary in Φ to $\langle +\text{fred}, +\text{citizen}\rangle$ and to

$$\langle +\text{fred}, +\text{citizen}, +\text{DISLIKES}[+\text{crook}], +\text{DISLIKES}[+\text{elected.crook}]\rangle.$$

Definition 5.4 Φ precludes a sequence $\langle x_1, \ldots, x_n, r[y_m], \ldots, r[y_1]\rangle$ iff there exists a pair $\langle z, r'[w]\rangle$ in Φ where z is an intermediary in Φ to $\langle x_1, \ldots, x_n\rangle$ and w is an intermediary in Φ to $\langle y_1, \ldots, y_m\rangle$.

Example: In the network of figure 5.3, Φ precludes the sequence

$$\langle +\text{fred}, +\text{citizen}, +\text{DISLIKES}[+\text{crook}], +\text{DISLIKES}[+\text{dick}]\rangle$$

because it contains the pair

$$\langle +\text{gullible.citizen}, -\text{DISLIKES}[+\text{elected.crook}]\rangle$$

and $+$gullible.citizen is an intermediary in Φ to $\langle +\text{fred}, +\text{citizen}\rangle$ and $+$elected.crook is an intermediary in Φ to $\langle +\text{dick}, +\text{crook}\rangle$.

Definition 5.5 A sequence $\sigma = \langle x_1, \ldots, x_n, r[y_m], \ldots, r[y_1] \rangle = \langle z_1, \ldots, z_p \rangle$ (where $m + n = p$) is *inheritable* in Φ iff Φ contains $\langle z_1, \ldots, z_{p-1} \rangle$, and either $\langle z_2, \ldots, z_p \rangle$ if $n > 1$ or $\langle y_1, \ldots, y_m \rangle$ if $n = 1$, and Φ neither contradicts nor precludes σ or σ^{-1}.

Example: If Φ contains $\langle +\text{clyde}, +\text{elephant} \rangle$ and $\langle +\text{elephant}, +\text{LOVES}[+\text{zookeeper}] \rangle$ then $\langle +\text{clyde}, +\text{elephant}, +\text{LOVES}[+\text{zookeeper}] \rangle$ is inheritable in Φ. If Φ consists of the two sequences $\langle +\text{elephant}, +\text{LOVES}[+\text{zookeeper}] \rangle$ and $\langle +\text{fred}, +\text{zookeeper} \rangle$ then

$$\langle +\text{elephant}, +\text{LOVES}[+\text{zookeeper}], +\text{LOVES}[+\text{fred}] \rangle$$

is inheritable in Φ.

Definition 5.6 Φ is *closed* under inheritance iff Φ contains every sequence inheritable in Φ.

Definition 5.7 Φ is an *expansion* of a set $S \subseteq \Sigma$ iff $\Phi \supseteq S$ and Φ is closed under inheritance.

Definition 5.8 Φ is *grounded* in a set $S \subseteq \Sigma$ iff every sequence in $\Phi - S$ is inheritable in Φ.

Definition 5.9 Φ is a *grounded expansion* of a set $S \subseteq \Sigma$ iff Φ is an expansion of S and Φ is grounded in S.

5.16   General theorems

Theorem 5.1 If Φ is a grounded expansion of Γ, then $\langle x_1, x_2 \rangle \in \Phi$ iff $\langle x_1, x_2 \rangle \in \Gamma$.

Proof   If $\langle x_1, x_2 \rangle \in \Gamma$ then $\langle x_1, x_2 \rangle \in \Phi$ because Φ includes Γ. If $\langle x_1, x_2 \rangle \in \Phi$ then $\langle x_1, x_2 \rangle \in \Gamma$ because Φ is grounded and sequences of length two are not inheritable in Φ.
∎

Theorem 5.2 If Φ is a grounded expansion of Γ then every sequence of Φ is of form $\langle x_1, \ldots, x_n, r[y_m], \ldots, r[y_1] \rangle$, where $n \geq 1$ and $m \geq 0$, x_1 and y_1 are positive individual or predicate tokens, x_2 through x_{n-1} and y_2 through y_m are positive predicate tokens, x_n is a predicate token and is positive if $m > 0$, and r is a signed relation.

Proof By induction on the length of sequences in the expansion and the definition of inheritability. ∎

Theorem 5.3 If a grounded expansion of Γ contains $\langle x_1, \ldots, x_n, r[y_m], \ldots, r[y_1] \rangle$, then it also contains all sequences $\langle x_i, \ldots, x_j \rangle$ for $1 \le i < j \le n$, $\langle y_i, \ldots, y_j \rangle$ for $1 \le i < j \le m$, and $\langle x_i, \ldots, x_n, r[y_m], \ldots, r[y_j] \rangle$ for $1 \le i \le n$, $1 \le j \le m$.

Proof By induction on the length of sequences in the expansion and the definition of inheritability. ∎

5.17 Symmetry

Definition 5.10 A set S of sequences is *symmetric* iff $\sigma \in S$ implies $\sigma^{-1} \in S$.

From here on we consider only symmetric sets of inheritance assertions. Thus, we assume throughout that Γ is symmetric.

Theorem 5.4 Let Φ be a grounded expansion of Γ. If Φ precludes σ then Φ precludes σ^{-1}.

Proof Suppose Φ precludes a sequence $\sigma = \langle x_1, \ldots, x_n, r[y_m], \ldots, r[y_1] \rangle$. If $m = 0$ then $\sigma = \sigma^{-1}$, so Φ precludes σ^{-1}. Assume $m > 0$. By definition there exists an intermediary z to $\langle x_1, \ldots, x_n \rangle$ and an intermediary w to $\langle y_1, \ldots, y_m \rangle$ such that $\langle z, r'[w] \rangle$ is in Φ. Therefore this pair is in Γ. Since Γ is assumed to be symmetric, $\langle w, r\prime^{-1}[z] \rangle$ is also in Γ, so it is in Φ. Thus Φ precludes $\langle y_1, \ldots, y_m, r^{-1}[x_n], \ldots, r^{-1}[x_1] \rangle$. ∎

Theorem 5.5 Every grounded expansion of Γ is symmetric.

Proof Γ is assumed to be symmetric. Suppose by way of contradiction that the theorem is false. Then let Φ be an asymmetric grounded expansion of a symmetric Γ, and let

$$\sigma = \langle x_1, \ldots, x_n, r[y_m], \ldots, r[y_1] \rangle$$

be a shortest relational sequence in Φ such that Φ does not contain the inverse sequence

$$\sigma^{-1} = \langle y_1, \ldots, y_m, r^{-1}[x_n], \ldots, r^{-1}[x_1] \rangle.$$

By the definition of inheritability, Φ contains $\langle x_2, \ldots, x_n, r[y_m], \ldots, r[y_1] \rangle$ (if $n > 1$, else $\langle y_1, \ldots, y_m \rangle$) and $\langle x_1, \ldots, x_n, r[y_m], \ldots, r[y_2] \rangle$ (if $m > 1$, else $\langle x_1, \ldots, x_n \rangle$.) Since these sequences are shorter than σ, if Φ contains them it also contains their inverses. Again by the definition of inheritability, since σ is inheritable in Φ, Φ neither contradicts nor precludes σ^{-1}. Therefore σ^{-1} is inheritable in Φ, so Φ cannot be closed under inheritance if it does not contain σ^{-1}. ∎

5.18 Ordering relations

The $\prec$ ordering on elements of Π and Θ remains the same as in chapter 2, although now it ranges over relational predicates and relational tokens as well. The definition of this ordering is repeated below verbatim. The $\prec$ ordering on sequences, however, is revised to treat relational sequences in the desired manner.

Definition 5.11 Let x and y be elements of Π. Then $x \prec y$ (read x is *below y*) iff either $\langle +x, +y \rangle \in \Gamma$, or for some $z \in \Pi$, $x \prec z$ and $z \prec y$.

Corollary 5.1 The $\prec$ relation is a partial ordering iff Γ is IS-A acyclic.

Definition 5.12 Let x and y be tokens in Θ. Then $x \prec y$ iff the $\prec$ relation holds between the corresponding elements of Π.

Definition 5.13 Let $\sigma = \langle x_1, \ldots, x_n \rangle$ and $\tau = \langle y_1, \ldots, y_m \rangle$ be sequences in Σ. If neither σ nor τ contain any relational tokens, then $\sigma \prec \tau$ iff $x_{n-1} \prec y_{m-1}$. If σ contains no relational tokens but τ contains at least one such token, then $\sigma \prec \tau$. Otherwise $\sigma \nprec \tau$.

Corollary 5.2 If Γ is IS-A acyclic then the $\prec$ relation on sequences is a partial ordering.

5.19 Consistency

Definition 5.14 A set of sequences is *consistent* iff it contradicts none of its elements.

Example: Figure 5.5 shows an inconsistent network. In this figure the sequences $\langle +\text{elephant}, +\text{LOVES}[+\text{zookeeper}] \rangle$ and $\langle +\text{elephant}, -\text{LOVES}[+\text{zookeeper}] \rangle$ contradict each other, since one says that elephants do love zookeepers while the other says elephants do not love zookeepers.

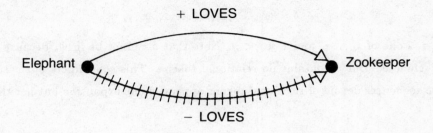

$$\Gamma = \Phi = \{\langle +\text{elephant}, +\text{LOVES}[+\text{zookeeper}]\rangle, \langle +\text{elephant}, -\text{LOVES}[+\text{zookeeper}]\rangle\}$$

Figure 5.5: An inconsistent network.

Theorem 5.6 A grounded expansion Φ of Γ is consistent iff Γ is.

Proof Same as proof of corresponding Theorem 2.8 in chapter 2. ∎

Theorem 5.7 Every union of distinct grounded expansions of an IS-A acyclic Γ is inconsistent.

Proof The proof is similar to the one in chapter 2. Let Φ_a and Φ_b be two distinct grounded expansions of an i.a. Γ. Let y be a minimal token such that $\sigma = \langle x_1, \ldots, x_n \rangle$ is a sequence in one expansion but not the other, where $x_{n-1} = y$ and σ contains no relational tokens. If such a sequence exists then by the method of chapter 2 we can show that $\Phi_a \cup \Phi_b$ is inconsistent. Assume that no such sequence exists. Then let $\sigma = \langle x_1, \ldots, x_n, r[y_m], \ldots, r[y_1] \rangle$ be a sequence of minimal length, and let x_n and y_m be minimal tokens, such that σ contains at least one relational token and σ is in one expansion but not the other. Assume without loss of generality that σ is in Φ_b and not in Φ_a. Φ_a contains both $\langle x_1, \ldots, x_n \rangle$ and $\langle x_2, \ldots, x_n, r[y_m], \ldots, r[y_1] \rangle$ if $n > 1$ and both $\langle y_1, \ldots, y_m \rangle$ and $\langle x_1, \ldots, x_n, r[y_m], \ldots, r[y_2] \rangle$ if $m > 1$. Then, since σ is not in Φ_a, Φ_a must either contradict or preclude it. If Φ_a contradicts σ the union of the two expansions is inconsistent. If Φ_a precludes σ, then $\langle x_1, \ldots, x_n \rangle$ and $\langle y_1, \ldots, y_m \rangle$ have intermediaries z and w, respectively, such that Γ contains a pair $\langle z, r'[w] \rangle$. Note that either $z \neq x_i$ for $1 \leq i \leq n$ or $w \neq y_j$ for $1 \leq j \leq m$, since otherwise every expansion would preclude σ.

157

Assume without loss of generality that $z \neq x_i$ for $1 \leq i \leq n$. Then Φ_a must include a sequence

$$\delta = \langle x_1, \ldots, x_i, z_1, \ldots, z_p, x_{i+1} \rangle$$

where z is one of $z_1 \ldots z_p$ and $1 \leq i < n$. Note that δ cannot be in Φ_b because Φ_b does not preclude σ. But δ contains no relational tokens. This contradicts our assumption that no sequences devoid of relational tokens appear in one expansion but not the other. ∎

5.20   Existence

Theorem 5.8 Every IS-A acyclic Γ has a grounded expansion.

Proof   Let Γ be i.a. We construct a grounded expansion of Γ by successively adding in inheritable sequences in such a way that later additions do not contradict or preclude (and hence invalidate) prior sequences.

If $\phi \subseteq \Sigma$ is i.a., we define $a(\phi)$, the set of augmentations of ϕ, to be the sets consisting of ϕ plus a single minimal inheritable sequence. Stated precisely, we first consider the set of all sequences inheritable in ϕ but not contained in ϕ; we write this set $I(\phi)$. We define $I^*(\phi) \subseteq I(\phi)$ to be the set of minimal inheritable sequences under the $\prec$ ordering. Then we define $a(\phi)$ so that $\phi' \in a(\phi)$ iff $\phi' = \phi \cup \{\sigma\}$ for some $\sigma \in I^*(\phi)$.

We note that since Σ is finite and $\prec$ is a partial ordering on Σ, $a(\phi) = \emptyset$ iff $I(\phi) = \emptyset$. We also note without proof that $a(\phi)$ may easily be computed from ϕ.

A sequence of successive approximations is defined to be any sequence $\{\phi_i\}_{i=0}^{\infty}$ such that $\phi_0 = \Gamma$ and for each $i \geq 0$, $\phi_{i+1} \in a(\phi_i)$ if $a(\phi_i) \neq \emptyset$; otherwise $\phi_{i+1} = \phi_i$. We will show that $\Phi = \cup_{i=0}^{\infty} \phi_i$ is a grounded expansion of Γ, no matter what choices of augmentations and order are made in the successive approximations.

We first note that $\phi_i \subseteq \phi_{i+1} \subseteq \Phi$ for each i, thus justifying the description "successive approximation." In particular, every sequence of ϕ_i contradicted (precluded) in ϕ_i is contradicted (precluded) in ϕ_{i+1}. We also note that if $\phi_{i+1} = \phi_i$ then $\phi_i = \Phi$, since $\phi_{i+1} = \phi_i$ means that $a(\phi_i) = \emptyset$, and hence $\phi_i = \phi_{i+n}$ for all $n \geq 0$.

To see that Φ is finite, first recall that in chapter 2 we showed that there are only a finite number of derivable sequences other than those containing relations. Every relational sequence $\langle x_1, \ldots, x_n, r[y_m], \ldots, r[y_1] \rangle$ is composed of two ordinary sequences:

158

$\langle x_1, \ldots, x_n \rangle$ and $\langle y_1, \ldots, y_m \rangle$. Since there are only a finite number of these ordinary sequences, there can be only a finite number of relational sequences for each relation. Therefore, assuming the number of relations is finite, only a finite number of relational sequences are derivable. Since each augmentation adds a new sequence to the approximation, we conclude that only a finite number of successive approximations can be augmentations. This means that for some $i \geq 0$, $\phi_i = \phi_{i+1} = \Phi$, so $|\Phi| \leq |\Gamma| + i$. (In fact, $|\Phi| = |\Gamma| + j$, where j is the least i such that $\phi_i = \phi_{i+1}$.)

Φ is clearly closed under inheritance, since $\Phi = \phi_i$ means that $\phi_i = \phi_{i+1}$, hence $a(\phi_i) = \emptyset$, hence $I(\phi_i) = \emptyset$, hence $I(\Phi) = \emptyset$.

Now we show that successive augmentations do not invalidate prior ones. There are several steps to this. First, we show that if $\phi_{i+1} = \phi_i \cup \{\sigma\}$ then for all $\tau \in I^*(\phi_j)$, $j > i$, $\tau \nprec \sigma$. The proof is by contradiction. Suppose τ is a sequence of minimal length such that $\tau \prec \sigma$ and $\tau \in I^*(\phi_j)$ for some $j > i$. Then τ contains no relational tokens. Since $\tau \prec \sigma$ and $\sigma \in I^*(\phi_i)$, $\tau \notin I(\phi_i)$. But $\tau \in I^*(\phi_j)$, so τ cannot be contradicted or precluded in ϕ_i. Therefore ϕ_i contains at most one of the two subsequences of τ, call them τ_1 and τ_2, that must be present in ϕ_j for τ to be in $I(\phi_j)$. Both $\tau_1 \prec \tau$ and $\tau_2 \prec \tau$, and τ_1 and τ_2 are of length one less than τ. But then, for some k, $i < k < j$, $\phi_{k+1} = \phi_k \cup \tau_1$ or $\phi_{k+1} = \phi_k \cup \tau_2$. This contradicts our assumption that τ was a sequence of minimal length falsifying the assertion.

Second, we show that each sequence of ϕ_i contradicted in ϕ_{i+1} is contradicted in ϕ_i. If ϕ_{i+1} contradicts a sequence $\sigma \in \phi_i$ not contradicted in ϕ_i, then $\phi_{i+1} = \phi_i \cup \{\tau\}$ for some $\tau \in I^*(\phi_i)$ such that τ contradicts σ. But then ϕ_i would contradict τ, so τ could not be inheritable in ϕ_i.

Third, we show that each sequence of ϕ_i precluded in ϕ_{i+1} is precluded in ϕ_i. If ϕ_{i+1} precludes a sequence $\sigma = \langle x_1, \ldots, x_n \rangle \in \phi_i$ that is not precluded in ϕ_i, then there exists some y that is an intermediary to σ in ϕ_{i+1} but not in ϕ_i, such that Γ contains a sequence $\langle y, u \rangle$ causing ϕ_{i+1} to preclude σ. Since y must become an intermediary to σ in ϕ_{i+1}, we see that

$$\phi_{i+1} = \phi_i \cup \{\delta\}$$

where $\delta = \{\langle x_1, l < dots, x_j, z_1, \ldots, z_p, x_{j+1} \rangle\}$ and y is one of $z_1 \cdots z_p$ and $1 \leq j < n - 1$. But $\delta \prec \sigma$, so $\delta \notin I^*(\phi_i)$.

Fourth, we show that every $\sigma \in \phi_i - \phi_0$ is inheritable in ϕ_{i+1}. Let σ be an element of ϕ_i and inheritable in ϕ_i. Then ϕ_i contains the two subsequences of σ that make σ inheritable, which means ϕ_{i+1} contains them too. Also, ϕ_i neither contradicts nor precludes σ. From the second and third steps of this proof, we conclude that σ is not contradicted or precluded in ϕ_{i+1}. Thus if σ is inheritable in ϕ_i it must be inheritable in ϕ_{i+1}.

Finally, to show that Φ is a grounded expansion of Γ, we note that $\Gamma = \phi_0$. Since $\Phi \supseteq \phi_0$ and Φ is closed under inheritance, Φ is an expansion of Γ. Since every element of $\Phi - \Gamma$ is inheritable in some ϕ_i, and hence in Φ, we see that Φ is grounded in Γ. ∎

5.21 Other properties

Other properties of relational inheritance networks remain to be explored. These include the conditions under which a network has a unique grounded expansion, the bounds on the size and number of possible expansions of a network, and the structure of the predicate-lattice model for networks containing relations.

5.22 More examples of relational inheritance

Figure 5.6 illustrates the interaction of relations with IS-A hierarchy exceptions. Since Orville loves elephants, he loves royal elephants and Clyde. Wilbur, who loves gray things, also loves elephants, but we do not infer that he loves royal elephants or Clyde, because they are not gray.

What happens if we add the explicit assertion that Wilbur loves elephants, as in figure 5.7? Wilbur loved elephants before this addition, since elephants are gray. But now we will also infer that Wilbur loves Clyde and royal elephants. The seemingly redundant assertion that Wilbur loves elephants turns out not to be redundant at all. The assertion really means that Wilbur loves elephants simply because they are elephants; this gives him a reason to love Clyde and royal elephants even though they are not gray.

Figure 5.8 shows that elephants love mammals in general, but they don't love elephants, including themselves. Thus, Clyde and Ernie love mammals but they don't love themselves or each other.

Figure 5.9 shows that tax collectors love tax collectors (including themselves.) Tax collectors do not love tax cheats. Clyde is both a tax collector and a tax cheat. Does

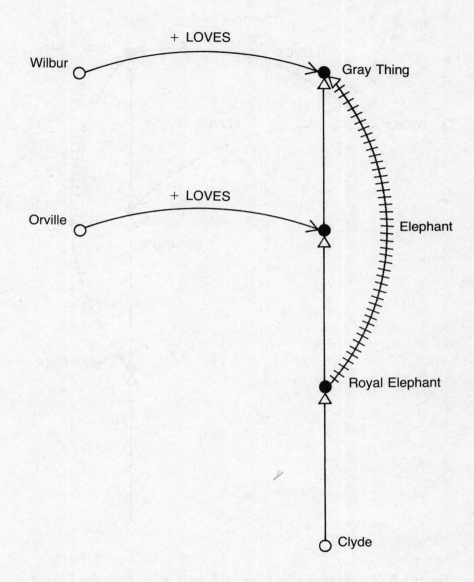

Figure 5.6: Wilbur loves gray things, while Orville just loves elephants.

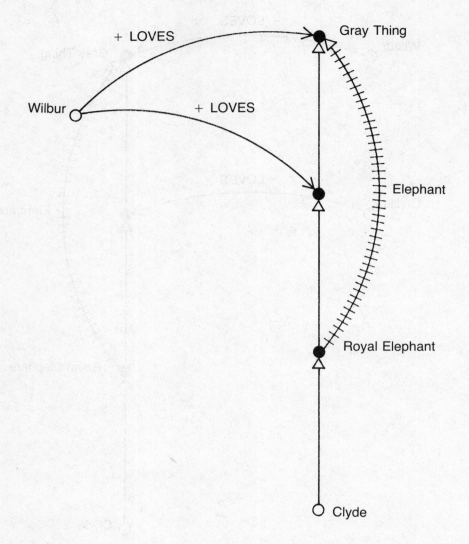

Figure 5.7: Wilbur loves gray things and independently he also loves elephants.

162

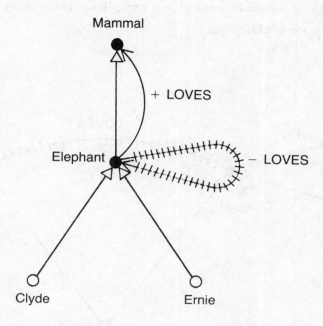

Figure 5.8: Elephants love mammals, except for elephants.

Clyde love himself, or not? This figure is ambiguous. In one expansion, Clyde loves himself; in the other he does not.

Figure 5.10 illustrates one way relational inheritance could be used to model default reasoning about event histories. The relation used is "before." There are two major classes of events in the example: those occurring as part of World War I, and those occurring during the early history of the Soviet Union. World War I mostly preceded the Russian revolution and the founding of the USSR. However, some events near the end of the war occurred after the beginning of the Russian revolution. By inheritance of the "before" relation between WW I events and early Soviet history events, we infer that Archduke Ferdinand was assassinated and the US entered the war before Czar Nicholas II's abdication, the end of the civil war in Russia, and Lenin's death. However, we see that Kaiser Wilhelm's surrender, which ended the war with Germany, occurred *after* the Czar had abdicated in Russia. We still infer by inheritance, though, that the Kaiser's surrender preceded both the end of the civil war and the death of Lenin, which in fact it did. The "before" relation we are using is not strictly transitive, since it is subject

163

to exceptions. A detailed theory of how exceptions should affect transitive relations is beyond the scope of this thesis.

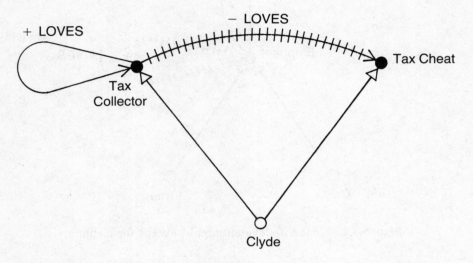

Figure 5.9: Tax collectors love tax collectors and hate tax cheats. Clyde is both a tax collector and a tax cheat.

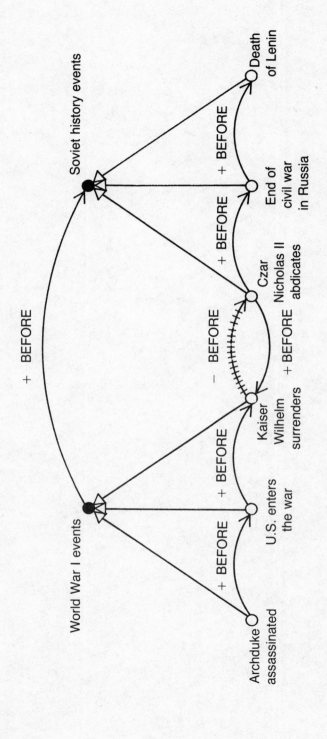

Figure 5.10: Using inheritance to model default reasoning about event chains.

165

6 Marker Propagation and Inheritable Relations

"All too often, serious work on representational issues in AI has been diverted or totally thwarted by premature concern for computational issues."
— Pat Hayes, *The Naive Physics Manifesto*

"So what's wrong with the computer? Part of the problem is that we do not yet fully understand the algorithms of thinking. But part of the problem is speed."
— Danny Hillis, *The Connection Machine*

6.1 Introduction

Having sketched a theory of inheritable relations in chapter 5, we now take up again the discussion of computational issues begun in chapter 4. In this chapter we will give a formal definition for the extension of a relational predicate, and then examine an algorithm, called the *relscan* algorithm, that reconstructs these extensions on a parallel marker propagation machine. The relscan algorithm is a combination of an upscan, a downscan, and a crossing of relational links. We will again find that marker propagation algorithms do not necessarily produce correct results on well-formed inheritance networks. Errors may occur in a relscan even when the network has been upscan and downscan conditioned. To assure the correctness of the relscan algorithm we must employ an additional preprocessing step known as relational conditioning.

6.2 Extensions of relational predicates

Relational predicates are written in abbreviated form as $R[y]$ or $R^{-1}[y]$, where R is a relation name and y is a positive predicate or individual token. The extension of the relational predicate $R[y]$ (for example, +LOVES[+fred], meaning "things that love Fred") can be written as a triple $\langle T_{Ry}, F_{Ry}, Q_{Ry} \rangle$ where, for a given expansion Φ,

$$T_{Ry} = \{x \mid \langle +x, +R[y]\rangle \in C(\Phi)\}$$

$$F_{Ry} = \{x \mid \langle +x, -R[y]\rangle \in C(\Phi)\}$$

$$Q_{Ry} = \Pi - (T_{Ry} \cup F_{Ry}).$$

Similarly, the extension of $R^{-1}[x]$ is written $\langle T_{xR}, F_{xR}, Q_{xR} \rangle$, and, for a given expansion Φ we have

$$T_{xR} = \{y \mid \langle +x, +R[+y] \rangle \in C(\Phi)\}$$

$$F_{xR} = \{y \mid \langle +x, -R[+y] \rangle \in C(\Phi)\}$$

$$Q_{xR} = \Pi - (T_{xR} \cup F_{xR})$$

We proved in chapter 5 that every grounded expansion Φ of a symmetric Γ is symmetric. The symmetry of Φ can be expressed in terms of extensions using the following equalities:

$$T_{xR} = \{y \mid x \in T_{Ry}\}$$

$$F_{xR} = \{y \mid x \in F_{Ry}\}$$

$$Q_{xR} = \{y \mid x \in Q_{Ry}\}$$

For the remainder of this chapter we will consider only relations of form $R[y]$; the corresponding inference algorithm for inverse relations $R^{-1}[x]$ is easily obtained by analogy.

6.3   The relscan algorithm

The algorithm we use for reconstructing relational extensions is, like the upscan and downscan algorithms, chosen because it is provably correct for certain restricted types of networks. It is not correct in the general case. Due to the faults of the downscan algorithm, the relscan algorithm is not even correct for systems which contain no exceptions or redundant links, as long as they permit general multiple inheritance and negative statements. No marker propagation scheme that uses shortest paths as a substitute for inferential distance can be correct in the general case. Later in this chapter I will give some networks where the relscan algorithm fails.

A relscan requires two inputs, a node x and a relation R. The algorithm reconstructs the extension of the relational predicate $R[x]$. The results are represented by markers

168

```
procedure relscan(x: node; R: relation; i, j: marker-triple) = begin
  clear[TM_i, FM_i, QM_i, TR_j, FR_j, QR_j];
  name[x]  ⟹  set[TM_i];
  loop      – – the upscan loop
    link-type[+R],
      on-tail[TM_i], off-head[TR_j, FR_j, QR_j]  ⟹  set-head[TR_j];
    link-type[−R],
      on-tail[TM_i], off-head[TR_j, FR_j, QR_j]  ⟹  set-head[FR_j];
    link-type[#R],
      on-tail[TM_i], off-head[TR_j, FR_j, QR_j]  ⟹  set-head[QR_j];
    link-type[IS-A],
      on-tail[TM_i], off-head[TM_i, FM_i, QM_i]  ⟹  set-head[TM_i];
    link-type[IS-NOT-A],
      on-tail[TM_i], off-head[TM_i, FM_i, QM_i]  ⟹  set-head[FM_i];
    link-type[NO-CONCLUSION],
      on-tail[TM_i], off-head[TM_i, FM_i, QM_i]  ⟹  set-head[QM_i];
  endloop;
  loop    – – the downscan loop
    link-type[IS-A],
      on-head[TR_j], off-tail[TR_j, FR_j, QR_j]  ⟹  set-tail[TR_j];
    link-type[IS-A],
      on-head[FR_j], off-tail[TR_j, FR_j, QR_j]  ⟹  set-tail[FR_j];
    link-type[IS-NOT-A, NO-CONCLUSION],
      any-on-head[TR_j, FR_j], off-tail[TR_j, FR_j, QR_j]  ⟹  set-tail[QR_j]
  endloop;
  off[TR_j, FR_j, QR_j]  ⟹  set[QR_j]
end
```

Figure 6.1: The relscan algorithm.

we shall call TR, FR, and QR. Internally, the relscan first performs an upscan of node x using another triple of markers which we shall call TM, FM, and QM. The complete relscan algorithm is given in figure 6.1.

The algorithm naturally divides into two halves, each of which is a loop. The first loop computes a partial upscan of x using markers TM, FM, and QM. (The reason the upscan is "partial" is that at the end the algorithm doesn't bother to mark all unmarked nodes with QM.) The upscan loop also sends TR, FR, and QR marks across relational links from nodes bearing TM. For example, if a node receiving marker TM has a $+R$ link from itself to some other node bearing none of TR, FR, or QR, then the latter node will be marked with TR the next time through the loop. Upon termination of the upscan half of the algorithm, the relational markers TR, FR, and QR will have been placed at various points in the IS-A hierarchy based on whatever relational links are present in the graph. Then, in the downscan loop, the TR, FR, and QR markers are propagated downward across IS-A links in order to find the complete extension of the relational predicate. Although it may seem strange during this part of the scan to place QR markers on the tails of IS-NOT-A and NO-CONCLUSION links that have TR or FR marks on their heads, this is necessary in order to handle IS-A hierarchy exceptions properly, as discussed in the next section.

Figure 6.2 gives an example of the algorithm operating on a simple network. The goal is to perform a relscan of Clyde with respect to the loves relation, *i.e.* find everything that Clyde loves. The algorithm begins by marking Clyde with TM_1, as shown in figure 6.2a. Then, the first time through the upscan loop, elephant is marked with TM_1, as shown in figure 6.2b. The second time through the upscan loop we notice the $+$LOVES link from elephant to zookeeper; since elephant bears TM_1, we send TR_1 across the link to zookeeper, as shown in figure 6.2c. The third time through the loop nothing happens, so the loop terminates. Now we enter the downscan loop and propagate TR_1 downward to find all the things Clyde loves. In this step Fred is marked with TR_1; see figure 6.2d. The second time through the loop nothing happens, so the loop terminates. Finally, in the last line of the algorithm, we complete the relscan by marking with QR_1 any nodes not already marked with one of TR_1, FR_1, or QR_1. The result is shown in figure 6.2e: Clyde loves zookeepers and Fred in particular, because these nodes bear TR_1; there is no conclusion about whether Clyde loves elephants or himself, which is indicated by their

170

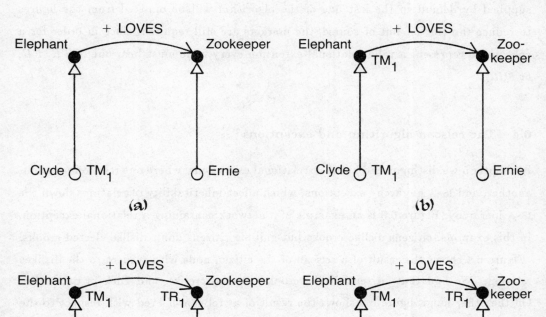

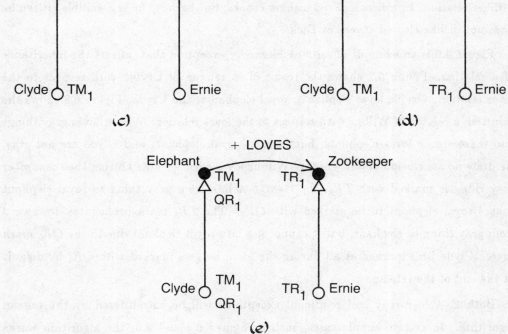

Figure 6.2: Steps in a relational scan.

bearing the marker QR_1. In the remaining examples in this chapter the QR markers supplied by default in the last line of the algorithm will be omitted from the figures to reduce the clutter, but of course, the markers are still required, since in order for a coloring to represent a valid relational extension every node must bear one of TR, FR, or QR.

6.4   The relscan algorithm and exceptions

In chapter 5 we distinguished between relational exceptions, where one relation overrides another, and IS-A hierarchy exceptions, which affect inheritability of relations down the IS-A hierarchy. Figure 6.3 is an example of a network containing a relational exception. In this example, citizens dislike crooks but gullible citizens don't dislike elected crooks. Figure 6.4 shows the result of a relscan of the citizen node with respect to the dislikes relation. We find that citizens dislike crooks, elected crooks, and Dick in particular. On the other hand, figure 6.5 shows the result of a relscan of Fred with respect to the dislikes relation. In this case, Fred dislikes crooks, but because he is a gullible citizen he does not dislike elected crooks or Dick.

Figure 6.6 is an example of an IS-A hierarchy exception that affects the inheritance of a relation. Figure 6.7 shows the result of a relscan of Orville with respect to the loves relation: Orville loves elephants, royal elephants, and Clyde. Figure 6.8 shows the result of a relscan of Wilbur with respect to the loves relation: Wilbur loves gray things and therefore he loves elephants, but because royal elephants and Clyde are not gray, we draw no conclusion about whether Wilbur loves them or not. During the scan, after gray thing is marked with TR_5, the IS-NOT-A link from gray thing to royal elephant causes royal elephant to be marked with QR_5. The TR_5 mark propagates downward from gray thing to elephant, but it cannot get into royal elephant due to the QR_5 mark there. Clyde isn't marked at all during the loop; he gets marked with QR_5 by default at the end of the relscan.

Both IS-A hierarchy and relational exceptions will be encountered by the relscan algorithm. In certain simple cases, such as figures 6.3 and 6.6, the algorithm works correctly. However, we shall soon see that where multiple inheritance and redundant IS-A links are involved, the algorithm is easily misled.

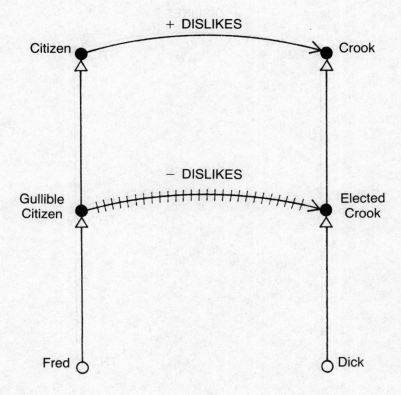

Figure 6.3: An example of a relational exception. Citizens dislike crooks, but gullible citizens do not dislike elected crooks.

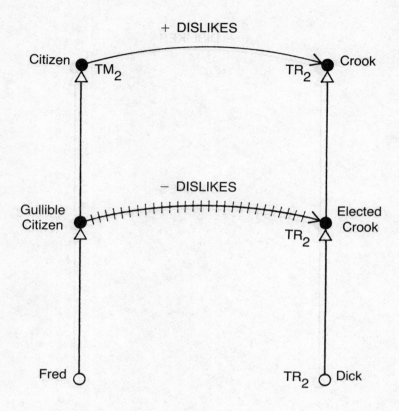

Figure 6.4: A relscan of citizen with respect to the DISLIKES relation. Citizens dislike crooks, elected crooks, and Dick. QR_2 marks have been omitted for clarity.

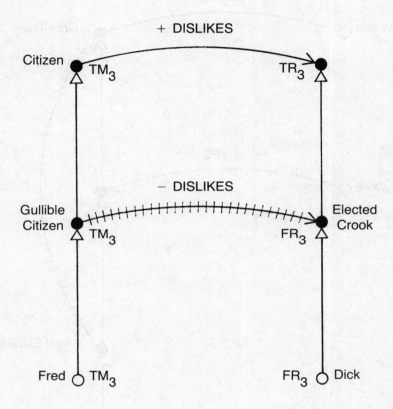

Figure 6.5: A relscan of Fred with respect to the DISLIKES relation. Fred dislikes crooks, but he does not dislike elected crooks or Dick. This is an example of a relational exception.

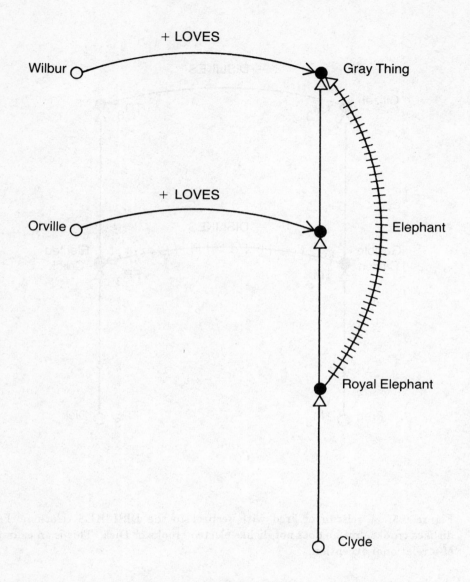

Figure 6.6: An example of an IS-A hierarchy exception that effects the inheritance of relations. The exception is that royal elephants are not gray things. Therefore lovers of gray things are lovers of elephants, but not necessarily of royal elephants. Such is the case with Wilbur.

176

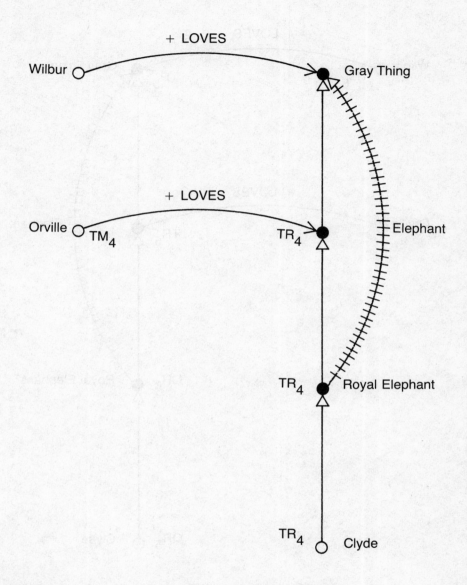

Figure 6.7: A relscan of Orville with respect to the LOVES relation.

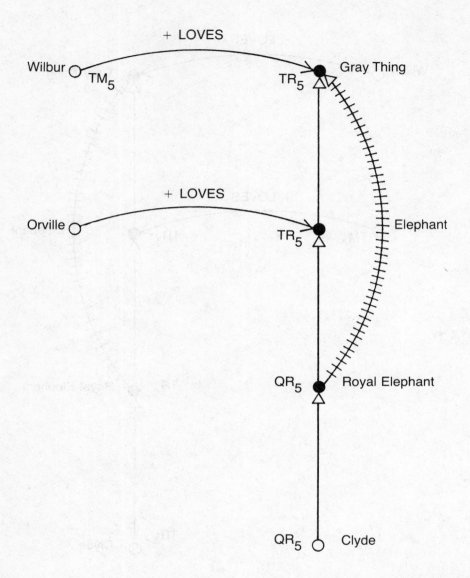

Figure 6.8: A relscan of Wilbur with respect to the LOVES relation. Wilbur loves gray things and elephants, but we draw no conclusion about whether he loves royal elephants or Clyde since they are not gray.

6.5   Correctness of the relscan algorithm

The relscan algorithm is correct for certain very restricted classes of networks. In chapter 4 we proved the correctness of two of its components, the upscan and downscan algorithms, for a class of networks called orthogonal class/property systems. In order to prove a scan algorithm correct we must pick a class of networks that have unique grounded expansions, since scan algorithms have no way to recognize ambiguity in a network or distinguish elements of the conclusion set of one expansion from those of another. A second requirement is that the network must be structured in such a way that the shortest path ordering always agrees with the inferential distance ordering.

When inheritable relations are considered, networks with extremely simple IS-A hierarchies can still be ambiguous. An example is figure 6.9. The IS-A hierarchy part of

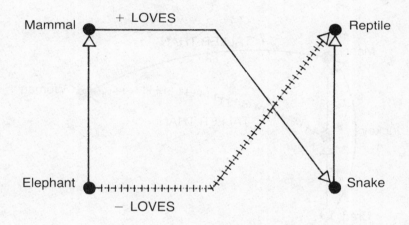

Figure 6.9: A minimal ambiguous network.

this figure qualifies as an orthogonal class/property system and is also exception-free. Yet figure 6.9 is ambiguous due to the crossed relational links.

Clearly we must further restrict the network topology when relational links are present in order to rule out ambiguity, but if we rule out exceptions entirely the resulting class of networks would be of little interest. Unfortunately there are a number of ways in which relational and IS-A hierarchy exceptions can interact to generate ambiguity. Thus, the classes of networks for which the relscan algorithm is correct must be small. I claim without proof that the relscan algorithm is correct on any network where the IS-A hierarchy is an orthogonal class/property system and the network contains at most one relational link at or above each node. (Figure 6.9 violates this constraint at both elephant and snake.) There are other classes of network for which the relscan algorithm is correct, but they are so restricted we will not attempt to enumerate them.

6.6 Places where the relscan algorithm fails

Two things may cause the relscan algorithm to fail, *i.e.* produce results not in agreement with the chosen expansion Φ. (Of course, if the network is unambiguous it will have only one expansion.) The first problem occurs in networks that contain redundant IS-A links. Redundant links act as shortcuts that can allow the wrong marker to win a race, thereby changing the result of the scan. An example appears in figure 6.10, where typically men are taller than women, but jockeys are typically men and are not taller

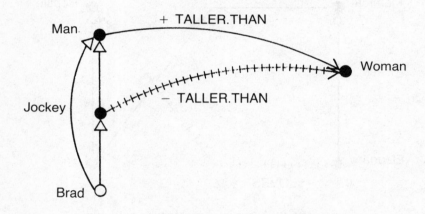

Figure 6.10: A relational network with a redundant IS-A link.

than women. Brad is a male jockey. The redundant IS-A link from Brad to man doesn't affect the upscan of Brad, but it does affect the relscan of Brad with respect to the taller-than relation. Figure 6.11 shows the result of the relscan algorithm. Jockey and man were marked with with TM_6 in the same iteration of the upscan loop; the algorithm then propagated TR_6 across positive relational links before sending FR_6 across negative ones. Brad was therefore wrongly inferred to be taller than the typical woman. The order in which the TR and FR marks are propagated is arbitrary; if we switch these

180

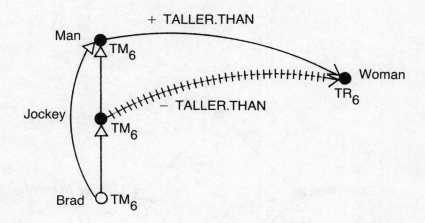

Figure 6.11: The relscan of Brad with respect to the TALLER.THAN relation, when run on this unconditioned network, incorrectly infers that Brad is taller than the typical woman. The problem is caused by the redundant IS-A link from Brad to man. Woman should have been marked with FR_6, not TR_6.

two lines in the algorithm we can construct a network similar to figure 6.10 in which again the "wrong" marker will be propagated first.

The second problem with the relscan occurs when the downward-propagating portion of the algorithm is forced to place a QR mark on some node prematurely. This is analogous to a problem we encountered with the downscan algorithm in chapter 4. The problem with ordinary downscans is illustrated by figure 6.12. In the figure, royal elephants are clearly drab things since they are shabby dressers. An upscan of the royal elephant node will indeed mark drab thing with TM. But a downscan of drab thing will mark royal elephant with QM, not TM, because after gray thing is marked with TM, a QM mark is send down the IS-NOT-A link that runs from royal elephant to gray thing. This network is obviously not downscan conditioned, and due to the inherent limitations of the downscan algorithm, we get the wrong result. A comparable situation for the relscan algorithm is shown in figure 6.13. Wilbur loves gray things, and Clyde is not gray. But Wilbur also loves performers, and Clyde is a performer. Therefore Wilbur loves Clyde. Unfortunately, the relscan algorithm, after marking gray thing and performer with TR during a relscan of Wilbur, will mark Clyde with QR due to the IS-NOT-A link. Clyde will therefore never receive TR if the relscan is run on the

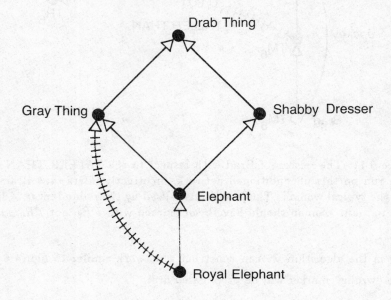

Figure 6.12: Royal elephants are clearly drab because they are shabby dressers. but a downscan of drab thing will mark royal elephant with QM rather than TM due to the IS-NOT-A link from royal elephant to gray thing. This network is not downscan conditioned.

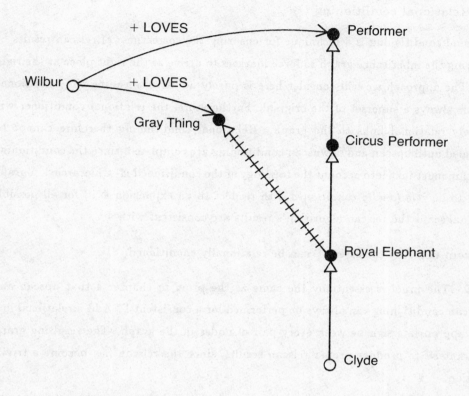

Figure 6.13: Wilbur loves gray things, and royal elephants are not gray. But Wilbur also loves performers. Royal elephants are performers, so Wilbur does love royal elephants. He therefore loves Clyde.

unconditioned network.

These two sources of failure for the relscan algorithm force us to develop additional conditioning measures in order to handle inheritable relations correctly.

6.7 Relational conditioning

Relational conditioning is a technique for ensuring the correctness of relscan results by modifying the inheritance graph to force markers to arrive at the right place at the right time. The approach we will consider here is purely additive, meaning the conditioned graph is always a superset of the original. Furthermore, the relational conditioner will add only relational links to the graph. Relational conditioning therefore cannot be performed until upscan and downscan conditioning are completed, since the conditioning algorithm must take into account the topology of the conditioned IS-A hierarchy. A graph is said to be *relationally conditioned* with respect to an expansion Φ iff for all possible relational scans the relscan algorithm's results are consistent with Φ.

Theorem 6.1 Any consistent Γ can be relationally conditioned.

Proof The proof is essentially the same as the proof in chapter 4 that upscan and downscan conditioning can always be performed for a consistent Γ. Add a relational link of the appropriate sign between every pair of nodes in the graph. The resulting graph is guaranteed to produce proper relscan results, since the relscan has become a trivial operation. ∎

This form of drastic relational conditioning does not even require the network to be upscan or downscan conditioned. However, it is woefully inefficient. A more efficient relational conditioning algorithm for i.a. networks is given below:

1. Determine the conclusion set $C(\Phi)$.

2. Let Γ' be an upscan and downscan conditioned network derived from Γ and $C(\Phi)$ by some conditioning method.

3. If there are no relations left to condition, stop. Otherwise, choose a relation R that has not been conditioned yet.

4. Condition Γ' with respect to relation R.

5. Condition Γ' with respect to R^{-1}, the inverse relation of R.

6. If any links were added in step 5, go back to step 4, else go to step 3.

The subprocedure for conditioning Γ' with respect to a particular relation R or its inverse R^{-1} is:

A. Pick a maximal node x (with respect to the $\prec$ ordering) that has not yet been examined. If all nodes have been examined, return.

B. Perform a relscan of node x with respect to the relation.

C. Pick a maximal node y (with respect to the $\prec$ ordering) whose relscan coloring disagrees with the correct coloring defined by $C(\Phi)$. If no such mis-colored y exists, return to step A.

D Add a relational link of the appropriate type from x to y in Γ', and return to step B.

TINA, my MacLisp program for constructing conclusion sets and conditioning IS-A acyclic networks, is also capable of relational conditioning. Here is an example of TINA conditioning the network in figure 6.10. The relation we are conditioning is named TALLER-THAN; the inverse relation is named SHORTER-THAN.

```
*(load-net 'jockey)                    ;Load the network shown
(C410DT50 JOCKEY TIN)                  ;in figure 6.10.


*lnet                                  ;List the assertions.
((LINK-5 : JOCKEY IS-A MAN)
 (LINK-4 : BRAD IS-A JOCKEY)
 (LINK-3 : BRAD IS-A MAN)
 (LINK-2 : MAN TALLER-THAN WOMAN)
 (LINK-2 : WOMAN SHORTER-THAN MAN)
 (LINK-1 : JOCKEY NOT-TALLER-THAN WOMAN)
 (LINK-1 : WOMAN NOT-SHORTER-THAN JOCKEY))
```

185

```
*(condition)                                    ;Invoke TINA.
Computing the expansion.
Conditioning the upscan.
Conditioning the downscan.
Conditioning the TALLER-THAN relation.
Added:  (BRAD NOT-TALLER-THAN WOMAN) to fix relational scan.
Conditioning the inverse relation, SHORTER-THAN.
Conditioning complete.
((BRAD NOT-TALLER-THAN WOMAN))                  ;Value returned is
                                                ;list of added links.


*lnet                                           ;Display revised network.
((LINK-6 : BRAD NOT-TALLER-THAN WOMAN)
 (LINK-6 : WOMAN NOT-SHORTER-THAN BRAD)
 (LINK-5 : JOCKEY IS-A MAN)
 (LINK-4 : BRAD IS-A JOCKEY)
 (LINK-3 : BRAD IS-A MAN)
 (LINK-2 : MAN TALLER-THAN WOMAN)
 (LINK-2 : WOMAN SHORTER-THAN MAN)
 (LINK-1 : JOCKEY NOT-TALLER-THAN WOMAN)
 (LINK-1 : WOMAN NOT-SHORTER-THAN JOCKEY))


*(plist 'brad)                                  ;One of TINA's internal
(SHORTER-THAN::MKRPATHS NIL                      ;data structures.
                SHORTER-THAN::INHPATHS
                NIL
                SHORTER-THAN::IMMED
                NIL
                TALLER-THAN::MKRPATHS
                (((JOCKEY +) (TALLER-THAN-WOMAN -)))
                TALLER-THAN::INHPATHS
```

```
                (((JOCKEY +) (TALLER-THAN-WOMAN -)))
                TALLER-THAN::IMMED
                ((TALLER-THAN-WOMAN -))
                RELDOWNS
                (((BRAD +)))
                IS-A-INHERITORS
                ((BRAD))
                MKRPATHS
                (((JOCKEY +))
                 ((MAN +))
                 ((JOCKEY +) (MAN +)))
                INHPATHS
                (((JOCKEY +)) ((JOCKEY +) (MAN +)))
                IMMED
                ((JOCKEY +) (MAN +))
                SIGN
                NIL
                TOPSORT
                O)
```

The conditioned version of the network that is the result of invoking TINA is shown in figure 6.14.

The network in figure 6.13 demonstrates how exceptions in the IS-A hierarchy cause problems for the downward-propagating portion of the relscan algorithm. Here is a demonstration of TINA conditioning this network. The relevant relation is LOVES; its inverse is named LOVED-BY.

```
*(load-net 'wilbur)                              ;Load the network show
(C410DT50 WILBUR TIN)                            ;in figure 6.13.


*lnet                                            ;List the assertions.
((LINK-5 : CLYDE IS-A CIRCUS-PERFORMER)
 (LINK-4 : CIRCUS-PERFORMER IS-A PERFORMER)
```

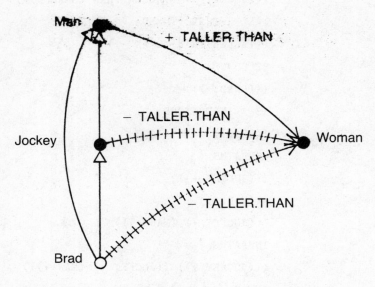

Figure 6.14: A conditioned version of figure 6.10.

```
(LINK-3 : CLYDE IS-NOT-A GRAY-THING)

(LINK-2 : WILBUR LOVES GRAY-THING)

(LINK-2 : GRAY-THING LOVED-BY WILBUR)

(LINK-1 : WILBUR LOVES PERFORMER)

(LINK-1 : PERFORMER LOVED-BY WILBUR))
```

```
*(condition)                                    ;Invoke TINA.
Computing the expansion.
Conditioning the upscan.
Conditioning the downscan.
Conditioning the LOVES relation.
Added:  (WILBUR LOVES CLYDE) to fix relational scan.
Conditioning the inverse relation, LOVED-BY.
Conditioning complete.
((WILBUR LOVES CLYDE))                          ;Value returned is
                                                ;list of added links.
```

188

```
*lnet                                        ;Display revised network.
((LINK-6 : WILBUR LOVES CLYDE)
 (LINK-6 : CLYDE LOVED-BY WILBUR)
 (LINK-5 : CLYDE IS-A CIRCUS-PERFORMER)
 (LINK-4 : CIRCUS-PERFORMER IS-A PERFORMER)
 (LINK-3 : CLYDE IS-NOT-A GRAY-THING)
 (LINK-2 : WILBUR LOVES GRAY-THING)
 (LINK-2 : GRAY-THING LOVED-BY WILBUR)
 (LINK-1 : WILBUR LOVES PERFORMER)
 (LINK-1 : PERFORMER LOVED-BY WILBUR))
```

The conditioned network output by TINA is shown in figure 6.15.

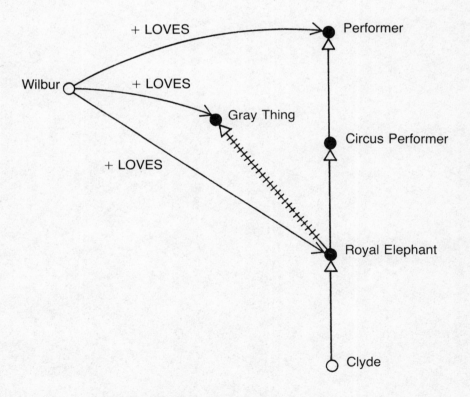

Figure 6.15: A conditioned version of figure 6.13.

6.8 Other considerations

As in chapter 4, we would like to know the the complexity of optimal or even reasonable conditioning algorithms, and the number of links that must be added to relationally condition a network. Worst case performance provides an upper bound, of course, but expected performance is a more relevant measure. Also, what is the cost of incrementally reconditioning a network after updating? These questions must be left as topics for future research.

7 Further Extensions to Inheritance

"What's the good of Mercator's North Poles and Equators,
Tropics, Zones, and Meridian Lines?"
So the Bellman would cry: and the crew would reply
"They are merely conventional signs!"

— Lewis Carrol, *The Hunting of the Snark*

"I look for a notation for things expressible, but unspoken."

— Frederick Parker-Rhodes, *Inferential Semantics*

7.1 Introduction

Opportunities abound for further research on inheritance systems and their formalizations. Some ways of extending the work presented in this thesis are discussed in the following sections.

7.2 Extending IS-A hierarchy inheritance

A simple but useful extension to the basic inheritance system would allow us to assert things about the complements of classes. In the inheritance language defined in chapter 2 this amounts to permitting negative predicate tokens to appear as the first elements of ordered pairs, *e.g.*

$$\langle -\text{elephant}, +\text{gray.thing} \rangle,$$

which says that non-elephants are gray. An interesting fact about sets forming a subset/superset chain is that their complements form a chain in the opposite direction, *e.g.* if elephant is a subset of mammal, then non-mammal is a subset of non-elephant. Since inheritance uses such chains of subset/superset relations, properties of the complements of sets would have to inherit *upward* along the IS-A hierarchy. For example, properties of non-elephants should be inherited by non-mammals. This change of direction will require some reworking of the inheritance axioms, and probably the notion of inferential distance as well, since now properties will inherit along both directions in the IS-A hierarchy.

Another area where the work in chapter 2 can be extended is the case of cyclic inheritance networks. Several theorems in chapter 2 have been proved only for IS-A acyclic networks. It would be useful to either extend the proofs to cover cyclic networks or construct some counterexamples.

7.3 Extensions to the inheritance of relations

If the basic inheritance system were extended to handle assertions about the complements of classes, *e.g.* ordered pairs of form $\langle -a, +b \rangle$, then we could proceed to extend relations to cover negative classes too, *e.g.*

$$\langle +a, +R[-b] \rangle.$$

This would allow us to represent relational statements such as "elephants love non-zookeepers." Note that the inverse of the above sequence is

$$\langle -b, +R^{-1}[a] \rangle.$$

The symmetry requirement, then, makes the latter extension dependent upon the former.

A completely independent extension would be to allow inheritance from relational classes, *e.g.* allow assertions of form $\langle +R[+a], +b \rangle$. This would enable us to represent statements such as "lovers of elephants are jolly." We could include negative relations such as $\langle -R[+a], +b \rangle$ (*e.g.* those who don't love elephants are jolly) with no extra effort, but assertions about the complements of classes, *e.g.* $\langle +R[-a], +b \rangle$ (lovers of non-elephants are jolly) would depend on the aforementioned extension to the basic inheritance system, due to symmetry.

7.4 Properties of relations

An entirely different line of extensions has to do with axioms for handling special properties of relations. In symmetric relations, R and R^{-1} are the same; these might be represented graphically by double-headed arrows. In antisymmetric relations, $+R$ is the same as $-R^{-1}$ and $+R^{-1}$ is the same as $-R$. In reflexive relations, $\langle +a, +R[+a] \rangle$ holds for all individual tokens. Similarly, for antireflexive relations, $\langle +a, -R[+a] \rangle$ holds for all individuals.

192

Currently there is no way to express the fact that a relation is reflexive over a certain domain without writing out the relations explicitly for every element. For example, if elephants love themselves, then the loves relation is reflexive over the domain of elephants. The sequence

$$\langle +\text{elephant}, +\text{LOVES}[+\text{elephant}]\rangle$$

does not express this idea; rather, it indicates that every elephant loves every elephant, including himself. In order to express reflexivity within the inheritance language (as opposed to just declaring the loves relation to be reflexive everywhere, using some means outside of the language of ordered pairs) we would have to add some sort of quantification operator.

Depending on whether exceptions are permitted, we might have strictly or weakly reflexive relations. To define strictly reflexive relations R, we simply state that for every individual a, the sequence $\langle +a, +R[+a]\rangle$ is an element of every grounded expansion. With weakly reflexive relations, $\langle +a, +R[+a]\rangle$ should appear in $C(\Phi)$ iff $\langle +a, -R[+a]\rangle$ and $\langle +a, \#R[+a]\rangle$ do not. Weakly reflexive relations permit exceptions to reflexivity, while strictly reflexive relations do not.

To define transitivity within the language would, like reflexivity, require a quantification operator, although we could build the axioms into the inheritance system directly and then simply declare a relation to be transitive. Transitive relations may also come in strict and weak forms. For strictly transitive relations R (*i.e.* those where transitivity is expressed by an ordinary implication rather than a default rule), if $\langle +a, +R[+b]\rangle$ and $\langle +b, +R[+c]\rangle$ are in $C(\Phi)$ then $\langle +a, +R[+c]\rangle$ must be in $C(\Phi)$ as well. See figure 7.1. This is a very powerful extension. For example, see figure 7.2, a network which is ambiguous. In one expansion we have $R(a, b)$ and $R(b, c)$, hence $R(a, c)$. In the other expansion we have $\sim R(a, c)$ and $R(b, c)$, hence $\sim R(a, b)$. Note that there is no $-R$ link between a and b or between any nodes above them. Therefore, in the second expansion we must conclude by some means other than pure inheritance that $\sim R(a, b)$ holds. An inheritance system that combined transitive relations with negation would be approaching predicate logic in expressive power. Such a system would bear a closer resemblance to a theorem prover than to a typical graph-traversing inheritance reasoner.

The weak form of transitivity permits exceptions to transitive relations. In figure 7.3, for example, from $R(a, b)$ and $R(b, c)$ we conclude $R(a, c)$ by transitivity, but

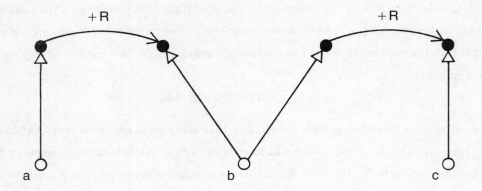

Figure 7.1: Example of inheritance of a transitive relation. Since $R(a,b)$ and $R(b,c)$ both hold, we conclude $R(a,c)$.

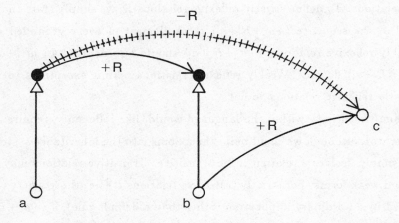

Figure 7.2: An ambiguous network. In one expansion we have $R(a,b)$, $R(b,c)$, and $R(a,c)$. In the other expansion we have $\sim R(a,c)$, $R(b,c)$, and $\sim R(a,b)$.

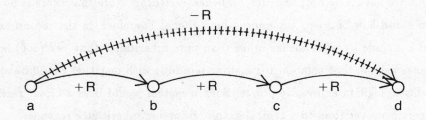

Figure 7.3: An example of weak transitivity. We conclude $R(a,c)$ but $\sim R(a,d)$.

since $\sim R(a, d)$ is explicitly asserted, from $R(a, c)$ and $R(c, d)$ we do not derive $R(a, d)$. According to the strict form of transitivity, figure 7.3 is inconsistent. Weak transitivity would seem to be a useful thing to have. The event chain network in figure 5.10 is an example of a weakly transitive relation: Most World War I events occurred before most events in the history of the USSR, but there is some overlap, so "before" cannot be strictly transitive if it is to be inheritable. Another weakly transitive relation is nearness: if Clyde's house is near Ernie's house, and Ernie's house is near Fred's house, then Clyde's house is probably near Fred's house. But perhaps not. If nearness were strictly transitive, everything would be near everything else if the nearness subgraph was connected.

Returning to figure 7.2, if we use weak transitivity we can get a grounded expansion in which $R(a, b)$, $R(b, c)$, and $\sim R(a, c)$ all hold. Adding weak transitivity to the inheritance system might be a less powerful extension than strong transitivity. It is not immediately obvious how weak transitivity should be defined, only that the axioms we decide upon will have to contain some notion of inferential distance. Of course, IS-A is a transitive relation subject to exceptions, but it's a special case. All other transitive relations inherit with respect to the IS-A relation, while IS-A itself does not depend on any other relation.

The last property of relations we will consider is that of being single-valued, *i.e.* if $R(a, b)$ holds then $R(a, c)$ does not, for all $c \neq b$. Such relations are sometimes called functional relations, since they are equivalent to functions. One application of functional relations is shown in figure 7.4. Here, elephants have color gray, and royal elephants have color white. We would like to conclude that royal elephants are not gray, rather than that they are both white and gray simultaneously. In a system without functional relations, we would have to put in a $-$COLOR link from royal elephant to gray to accomplish this. With functional relations, though, the inference that royal elephants are not gray falls out automatically. Slots in FRL are essentially functional relations, in that specifying a value for royal elephant's color slot prevents its inheriting elephant's value for that slot.

If the color relation were treated as functional in figure 7.5, this network would have three expansions. In one of them, Clyde, Ernie, and elephants in general would be light gray; in one they would be medium gray; and in one they would be dark gray. In contrast, if color is treated as an ordinary relation according to the rules of chapter 5

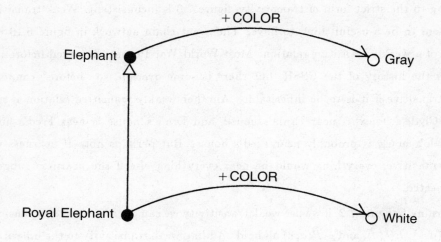

Figure 7.4: An application of functional relations. Since royal elephants are white we should infer that they are not gray, because the COLOR relation is functional.

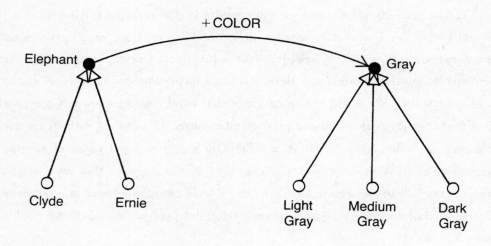

Figure 7.5: A network with several possible interpretations.

(*i.e.* doubly universally quantified, modulo exceptions) then Clyde, Ernie, and elephants in general would be all three shades of gray at once. The most likely intended meaning of figure 7.5 is that elephants are typically *some* shade of gray. To represent this accurately would require some form of existential quantification.

7.5 A hierarchy of relations

Fahlman has suggested organizing relations themselves into a taxonomic hierarchy. Then, from $\langle +R, +S \rangle$ and $\langle +a, +R[+b] \rangle$ we could derive $\langle +a, +S[+b] \rangle$. For example, if R is the relation "major part of" and S is the relation "part of," then $\langle +R, +S \rangle$ and $\langle -S, -R \rangle$ hold. The relational hierarchy, since it is an application of the IS-A hierarchy, could involve multiple inheritance and exceptions — an intriguing possibility.

One way to declare relational properties from within the inheritance system would be to admit statements such as

$$\langle +R, +transitive.rel \rangle$$

where $+transitive.rel$ is a distinguished predicate implying the applicability of a set of axioms for transitivity. (Note that the properties of $+R$ are not necessarily shared by $-R$. For instance, if R is the $=$ relation then $+R$ is transitive but $-R$ is not. On the other hand, if R is the arithmetic $<$ relation then both $+R$ and $-R$ are transitive. If R is the color relation than $+R$ might well be functional but $-R$ can be functional only when no two objects in the world have the same color.) From a logician's viewpoint, declaring properties of relations this way may seem strange. But it is useful for an inheritance reasoner to be able to describe itself in itself. One could imagine describing the color relation with assertions such as the following:

$$\langle +COLOR, +functional.rel \rangle$$

$$\langle +COLOR, +DOMAIN[+physob] \rangle$$

$$\langle +COLOR, +RANGE[+color.value] \rangle$$

and then using the relation by writing, say,

$$\langle +clyde, +COLOR[+gray] \rangle.$$

7.6 Relations of higher arity

Yet another possibility for extending the theory of inheritable relations is to generalize from binary to n-ary relations. We do this by converting n-ary relations to unary predicates through repeated lambda abstractions. The idea is to represent the relation $R(x_1, \ldots, x_n)$ by a family of n derived predicates written in shorthand form as

$$R[\bullet, +a_2, \ldots, +a_n] \quad \cdots \quad R[+a_1, \ldots, +a_{n-1}, \bullet].$$

Here, the $\bullet$ symbol is used to explicitly mark the position of the lambda variable. Figure 7.6 shows an inheritable ternary relation "x prefers y to z." In this example, elephants prefer cars to motorcycles, but elephants do not prefer compact cars to Fred's Harley. N-ary relational statements are represented in graphical form by relation nodes with numbered arrows indicating the positions of the arguments. In sequence form the two relational statements in figure 7.6 would each be represented by three ordered pairs:

$\langle +\text{elephant}, +\text{PREFERS}[\bullet, +\text{car}, +\text{motorcycle}] \rangle$

$\langle +\text{car}, +\text{PREFERS}[+\text{elephant}, \bullet, +\text{motorcycle}] \rangle$

$\langle +\text{motorcycle}, +\text{PREFERS}[+\text{elephant}, +\text{car}, \bullet] \rangle$

$\langle +\text{elephant}, -\text{PREFERS}[\bullet, +\text{compact.car}, +\text{freds.harley}] \rangle$

$\langle +\text{compact.car}, -\text{PREFERS}[+\text{elephant}, \bullet, +\text{freds.harley}] \rangle$

$\langle +\text{freds.harley}, -\text{PREFERS}[+\text{elephant}, +\text{compact.car}, \bullet] \rangle$

The definitions of intermediaries, preclusion, and inheritability in chapter 5 would have to be rewritten to support inheritance of higher arity relations. At first glance this does not appear difficult to do, but the details have not actually been worked out.

Unfortunately it does not appear possible to do parallel inheritance of higher arity relations on a parallel marker propagation machine. In NETL, Fahlman realized that relational statements (called *IST nodes, for "individual statements") would have to be handled one at a time by the inference algorithm.

7.7 Quantification

The prime reason for omitting explicit quantification from inheritance systems is that it would require a much more complicated inference algorithm. Simple, graph-traversal

198

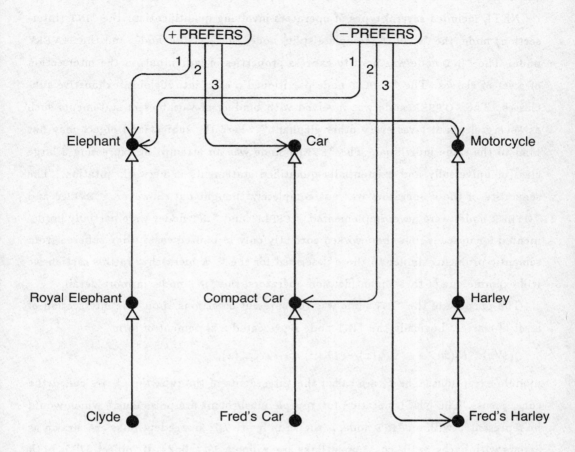

Figure 7.6: An example of a ternary relation. Elephants prefer cars to motorcycles. But elephants do not prefer compact cars to Fred's Harley. The syntax of the ternary relation is "1 prefers 2 to 3."

inference algorithms are one of the major attractions of inheritance systems. Another problem, though, is that there may be some semantic difficulties with quantification when exceptions are present. This is an interesting question and merits further study.

NETL included several types of operators involving quantification: the *INT (intersection) node, the *CSPLIT (complete split) node, the *OTHER node, and the *EVERY node. The *INT node was used to express properties of individuals in the intersection of a set of classes. The *CSPLIT node partitioned a class into disjoint, exhaustive subclasses. The *OTHER node was involved with binding of variables in statements such as "every elephant loves every other elephant," where the subject and object may not bind to the same individual. The *EVERY node was an attempt at expressing a large class of universally and existentially quantified statements in network notation. The semantics of these operators was not completely thought out, however. *EVERY and *OTHER nodes were never implemented. *CSPLIT and *INT nodes were partially implemented for upscans, but they worked correctly only in limited cases; they suffered from semantic problems similar to those described for the IS-A hierarchy. In this section we will examine one of these quantification operators, the *INT node, in more detail.

The purpose of the *INT node was to represent assertions about the intersection of a set of classes. Logically the *INT node represented a statement of form

$$(\forall x) \quad P_1(x) \wedge \cdots \wedge P_n(x) \rightarrow Q_1(x) \wedge \cdots \wedge Q_m(x),$$

modulo exceptions. The P_i are called the antecedents of the rule; the Q_i are called the consequents. The NETL notation for "purple mushrooms are poisonous," which would be represented with an *INT node, is shown in figure 7.7; antecedent links are drawn as arrows with barbs, while consequent links are ordinary IS-A links. If "object 37" is both a mushroom and a purple thing, the intent is that we infer that it is poisonous. But suppose we add a rule that *small* purple mushrooms are not poisonous, as in figure 7.8. We would need some way for the inferential distance ordering to come in and allow the second rule to act as an exception to the first. The details of exactly how this should be done have yet to be worked out.

*INT nodes provide a new way in which networks can be made ambiguous. For example, if purple mushrooms are poisonous and spotted mushrooms are not poisonous, what conclusion should we draw about a spotted purple mushroom? Probably such a network should have two grounded expansions.

200

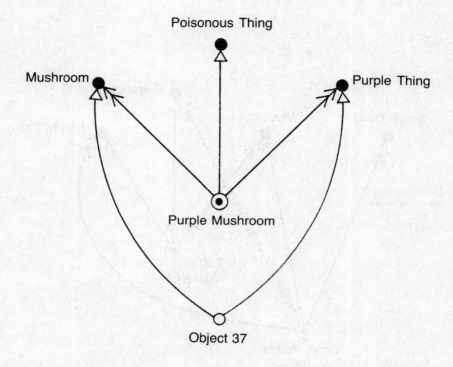

Figure 7.7: An example of an *INT node in NETL. The *INT node says that purple mushrooms are poisonous. Since object 37 is purple and is a mushroom, we should infer, due to the *INT node, that it is poisonous.

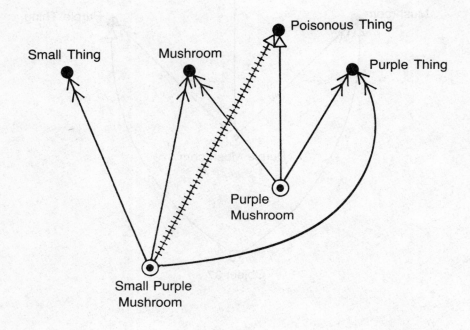

Figure 7.8: An example of an *INT node exception. Purple mushrooms are poisonous, but small purple mushrooms are not poisonous.

7.8 Other applications of exceptions

Exceptions occur in other areas of knowledge representation besides the IS-A hierarchy. Consider the part-of hierarchy. The part-of relation, which inherits along IS-A links, is subject to exceptions such as "cars have as parts gas tanks, except that electric cars do not." An entirely different form of part-of exception, suggested by Pat Hayes (personal communication) is an exception to the properties that inherit along part-of links. For example, cars are made of metal. Car bodies are parts of cars, so by inheritance they are made of metal. Car bumpers are parts of car bodies, but these days they are usually made of plastic rather than metal. See figure 7.9. Now, consider the front bumper of a

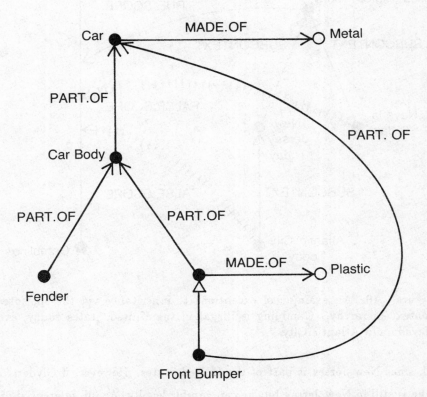

Figure 7.9: An example of an exception in the part-of hierarchy. Cars are made of metal, but bumpers, which are parts of cars, are made of plastic.

car. If we state explicitly that it is part of the car, a naive inheritance reasoner might infer it is made of metal, but on the other hand, since it is a bumper, it is probably plastic. Clearly some notion of inferential distance is needed to assure that exceptions are handled properly in the part-of hierarchy.

Another place where exceptions occur is the scope (*i.e.* spatiotemporal context) in which an assertion holds or does not hold. Fahlman gives the following example: casino gambling is illegal in the United States today, but it is not presently illegal in Nevada or Atlantic City. The latter two contexts are subcontexts of "the United States today." See figure 7.10. If Clyde is in New Jersey and is gambling, then he is doing something

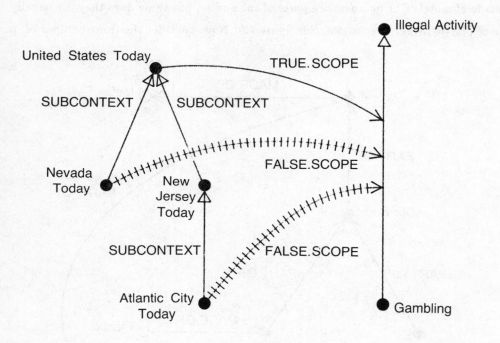

Figure 7.10: An example of exceptions to inheritance via the spatiotemporal context hierarchy. Gambling is illegal in the United States today, except in Nevada and Atlantic City.

illegal, since New Jersey is part of the United States. However, if Clyde is in Atlantic City, he is still in New Jersey but he can gamble legally. Again, inferential distance can be used to assure that exceptions are handled correctly in the presence of redundant statements such as "Clyde is in New Jersey."

The subcontext relation is an interesting one, because the scope of a subcontext relation can be limited to another subcontext. In figure 7.11, Hawaii is shown to be a subcontext of the United States. But while this subcontext relation holds in the context "the 20th century," it does not hold in the context "the 18th century." Or consider the sentence "in 1975 it became illegal to sell any motorcycle to an elephant after 1978; this law was suspended for one year in 1983." The resulting world state is shown in figure 7.12. Note the similarity between the structure of this diagram, which contains a scope exception, and the structure of a network containing a relational exception.

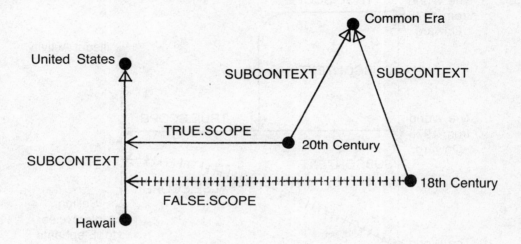

Figure 7.11: An example in which the SUBCONTEXT relation is itself true only in a certain subcontext. Hawaii is a subcontext of the United States in the 20th century but not in the 18th century.

A theory of inheritance reasoning across spatiotemporal contexts with various sorts of exceptions would be a significant extension to this thesis. Some of the basic questions that must be answered are the conditions under which a network is consistent, the conditions under which it is unambiguous, the conditions under which it has a grounded expansion, and the proper definition for inferential distance with respect to scope and contexts.

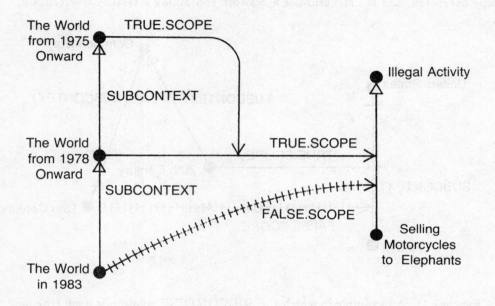

Figure 7.12: In 1975 it became illegal to sell motorcycles to elephants after 1978; this law was suspended for one year in 1983.

8 Conclusions

> "...it has proved, in the long run, to be just as amenable to pictorial, non-mathematical expression; it has given the right answers to some decisive questions; it is the theory most theoreticians prefer."

— Ian Fleming, on molecular orbital theory

8.1 Historical summary

In the beginning there were taxonomic hierarchies. Inheritance reasoning was fast because it was simply transitive closure. And AI hackers saw that this was good.

Then came frame systems with defaults and exceptions. Inheritance reasoning became nonmonotonic, which led many to believe that it was somehow outside of logic, until Hayes came to logic's defense, and McDermott and Doyle and Reiter showed how logic could be nonmonotonic. But while general purpose logical theorem provers were slow and nonmonotonic logic wasn't even decidable, inheritance reasoners were fast because they used a simple path length ordering to deal with exceptions. Shortest path reasoning was exactly the right thing for dealing with exceptions then, because the shortest inference path was always the correct one. And AI hackers saw that while inheritance was good, inheritance with exceptions was even better.

Then came multiple inheritance. And with multiple inheritance came a multiplicity of inheritance algorithms. Some continued to use shortest path reasoning. Others searched all possible inheritance paths. And some had unpredictable behavior. But all of the algorithms were fast, even if their behavior didn't always make sense. And AI hackers saw that that was okay: their inference algorithms did "the right thing" in most "reasonable" cases, and that was enough to get work done. Besides, nobody wanted to go back to using logic, especially the undecidable kind.

Finally there came the inferential distance ordering. At last "the right thing" had a formal definition. And inheritance systems now could do "the right thing" all of the time. But alas, the right thing involves only a partial ordering, so that sometimes to do the right thing one must report an ambiguity instead of a definite answer. And alas, it takes longer to do the right thing (by computing inferential distance) than to search blindly up an IS-A chain and grab the first answer one finds, as shortest path reasoners do. And alas, although parallel marker propagation machines can compute transitive closures quickly, there is no known way for them to handle inferential distance unless the network is conditioned in advance. Oh, woe to the AI hacker who insists that a system do the right thing all of the time!

8.2   Inferential distance in 25 words or less

The essential intuition behind inheritance exceptions is: subclasses override superclasses. Briefly stated, the inferential distance rule says that A may view B as a subclass of C iff A has an inference path *via* B to C, and not vice versa. Suppose we are trying to find out if A has property P. If A inherits from B which has property P, and also from C which has property $\sim P$, what conclusion should we reach about A? The inferential distance ordering says: if A has an inference path via B to C and not vice versa, then conclude that A has P; if A has an inference path via C to B and not vice versa, then conclude that A has $\sim P$; otherwise there is an ambiguity.

8.3   Results of the thesis

AI hackers want inheritance reasoners to do "the right thing." In this thesis I say clearly what the right thing is. This had never been done before for inheritance systems that admit exceptions. In the past, most inheritance systems were constructed in an *ad hoc* manner with no mathematical investigation of their structure. (But see (Attardi and Simi, 1981) for a formal treatment of inheritance systems that do not permit exceptions.) As a consequence, semantic problems lay undiscovered in systems with both multiple inheritance and exceptions. One problem was an inability to handle redundant statements reasonably. Another was an inability to recognize ambiguity in a set of assertions. The notion of inferential distance, which I formally defined in this thesis, offers a solution to both these problems.

In this thesis I also defined a formal language in which algorithms for parallel marker propagation machines can be concisely expressed. I then presented a set of PMPM algorithms for inheritance reasoning and showed that they were correct only in certain

208

limited cases. (Note that a discussion of correctness is only possible when one can distinguish what one wants to do from how one proposes to do it. Knowledge representation theories expressed solely in terms of their implementations do not meet this condition.) A fundamental limitation of PMPM's was uncovered: since they do not appear capable of computing inferential distance, their inheritance algorithms are correct only on networks where the path length ordering agrees everywhere with the inferential distance ordering. Even in networks with no redundant links and no exceptions, the downscan and relscan algorithms are not guaranteed to work correctly. The upscan and downscan algorithms were proved correct for one useful class of networks, though: orthogonal class/property networks.

One way to get around the limitations of PMPM's is to modify the network by adding or removing links to ensure that one's PMPM algorithms produce the desired results. This technique was given the name "conditioning." I explored this possibility in detail in the thesis, first proving that conditioning is always possible, then giving a simple but reasonably efficient conditioning algorithm.

Finally, I extended the concept of inheritance by introducing a useful new piece of knowledge representation machinery: inheritable relations with exceptions. I gave formal definitions showing how such relations should treat relational exceptions and how they should interact with the IS-A hierarchy and its own class of exceptions. The implementation of inheritable relations on a PMPM was also discussed.

8.4 Summary of major theorems

The major theorems of the thesis are summarized below. Γ denotes an inheritance network.

Theorem 2.4 No grounded expansion is a subset of another.

Corollary 2.2 Grounded expansions are minimal expansions.

Theorem 2.7 Every grounded expansion of an IS-A acyclic Γ is finite.

Theorem 2.8 A grounded expansion of Γ is consistent iff Γ is.

Theorem 2.11 Every IS-A acyclic Γ has a constructable grounded expansion.

Theorem 2.13 If a totally acyclic Γ is ambiguous then each of its grounded expansions is unstable.

Theorem 2.14 If a totally acyclic Γ has an unstable grounded expansion then it has

209

more than one grounded expansion.

Corollary 2.9 A totally acyclic Γ is unambiguous iff it has a stable grounded expansion.

Theorem 4.2 If Γ is consistent then it is conditionable.

8.5 The relationship between inheritance systems and logic

How does an inheritance system differ from a logic? If one adopts a suitably vague definition of "logic" then almost any formal system will qualify, so I will restrict the discussion to first order logic and nonmonotonic logic.

First of all, the language of inheritance systems is highly restricted. Assertions contain no conjunctions or disjunctions, and at most a limited form of negation. In contrast, FOL and NML permit expressions of arbitrary complexity. By adopting a suitably restricted language it is possible to reduce inference to a nearly trivial operation. Inference is not a trivial operation in FOL; in NML it is not even mechanizable. In contrast, most inheritance reasoners are just graph traversal algorithms, which are far simpler than, say, resolution theorem proving.

Secondly, inheritance systems have a taxonomic flavor. They are founded on the assumption that knowledge is structured hierarchically, and this assumption defines their style of nonmonotonic reasoning. NML makes no such assumption. NML can therefore handle other types of nonmonotonic reasoning, but because NML lacks the implicit notion of hierarchy, exceptions to NML inference rules must always be handled explicitly, while inheritance systems can handle exceptions implicitly. This problem was discussed in detail in chapter 1. Inheritance systems seem to provide a more natural, intuitive treatment of exceptions than NML, but only so long as the hierarchical assumption is not violated. Consider the following set of assertions:

Unemployed persons are typically adults.

Adults are typically employed.

Adults are typically high school graduates.

High school graduates are typically adults.

High school graduates are typically employed.

Employed persons are typically high school graduates.

There is no hierarchical structure to this example. The most we can say is that the classes adult, high school graduate, and employed person have a high degree of mutual overlap. Since this example involves cyclic inheritance paths, it cannot be represented in most inheritance systems. I have not outlawed cyclic structures in the generic system defined in this thesis, but several of the key theorems were proved only for the acyclic case and may not hold when the network contains cycles.

8.6 Inferential distance applied to default logic

Etherington and Reiter recently showed how inheritance systems can be formalized as a special case of default logic (Etherington and Reiter, 1983). They are able to prove certain theoretical results about inheritance systems as logical systems by doing so. But again, the problem with representing inheritance networks in default logic is that each rule must either mention exceptions explicitly, or else mention the names of rules to be defeated. This begs the questions of how exceptions should be treated.

It appears that we can escape from this difficulty by using inferential distance. The idea is to define an "application ordering" on default rules based on the topology of the inheritance graph (Touretzky, 1984). We then associate with each inference a proof sequence (analogous to an inheritance path) that lists the rules that justify that inference. We outlaw proof sequences in which the application ordering is violated. The result is that the links of an inheritance graph can be represented simply as normal defaults, even when exceptions are present. The parallels between this version of default logic and the generic inheritance system of chapter 2 are clear.

8.7 Usefulness of parallel marker propagation in AI

How good is parallel marker propagation as an architecture for AI? PMPM's were designed to provide fast set intersection and fast transitive closure, and at that they are successful. However, the inability of the PMPM to compute inferential distance leads us to question whether it can implement any more general set of knowledge base operations. Fahlman has noted that PMPM's cannot compute join operations, such as finding all fathers who hate their own sons (Fahlman, 1982). Joins would appear to be of fundamental importance in knowledge representation. Although certain trivial types of problems can be solved with set intersection and transitive closure alone, this does not seem to justify adopting the PMPM as an architecture for AI reasoning.

211

Fahlman wanted to answer a particular question which led him to marker propagation. Namely, how are people able to intersect a collection of features such as "big, gray, long-nosed quadruped" and come up with "elephant" as quickly as they do? As a psychological theory PMPM's have much in common with spreading activation theories (Quillian, 1968), (Anderson, 1983). They provide a plausible answer to a very intriguing problem. If people do turn out to do rapid feature intersection unconsciously by marker propagation, it is perhaps not unreasonable to suggest that they condition their internal inheritance networks through slower, conscious reasoning when exceptions are encountered.

Another parallel architecture which was created to implement the NETL theory of knowledge representation is Hillis' Connection Machine, which can perform a limited sort of parallel message passing (Hillis, 1985). Unfortunately it appears that the Connection Machine, although somewhat more powerful than a PMPM, may still be unable to do efficient inheritance reasoning using inferential distance. Further investigation is necessary.

8.8 Practical applications of the thesis

AI researchers favor inheritance representations because they are simple and efficient. Systems based on inferential distance will be less so. Therefore I do not expect system implementors, after reading this thesis, to rush out and convert their inheritance reasoners to use inferential distance. However, the research may still have some practical benefits for them. First, it explicitly defines the semantic problems faced by multiple inheritance systems: (1) redundant links can confuse inheritance reasoners that use a path length ordering rather than inferential distance, and (2) shortest path reasoners fail to recognize ambiguity. Now at least everyone knows that the problems exist. Second, although implementors may choose for efficiency reasons not to have their systems actually *do* the right thing in all cases, they at least have a formal theory to refer to telling them what the right thing is. Third, readers are reminded that there is a way to avoid semantic problems altogether at some cost of generality: use restricted network topologies such as orthogonal class/property systems.

212

8.9 Theoretical applications of the thesis

Users of inheritance systems find the implicit hierarchical ordering of defaults to be convenient. This is something that nonmonotonic and default logic systems presently lack. Having defined the right ordering to use for hierarchy-based reasoning, this ordering can now be incorporated into logic-based systems as *one component* of their representation of the world. (The Krypton system is an example of a general reasoner that contains an inheritance reasoner as a special component (Brachman, Fikes, and Levesque, 1983). But Krypton does not permit inheritance exceptions.) There may still be many occasions in commonsense reasoning where explicit handling of exceptions is necessary, but in those cases where an inheritance reasoner using the inferential distance ordering would handle an exception implicitly, now a default or nonmonotonic logic reasoner can too.

I can imagine two ways in which inferential distance might be applied in logical reasoning systems. One is to follow Etherington's suggestion (personal communication) that inferential distance act as a filter over the set of admissible expansions a default theory would generate. This, I feel, is a reasonable theoretical view. A practical approach that a logic-based reasoning program could take would be to use a preprocessor to transform sets of normal-form default rules into seminormal ones in accordance with the inferential distance ordering. This approach becomes more necessary as the knowledge base grows in size, since as the number of exceptions to a rule increases, so does the number of expansions to be filtered; likewise the complexity of the seminormal versions of the default rules increases.

8.10 Fahlman's virtual copy idea

One of the key ideas in NETL was that of virtual copies, which Fahlman describes below (Fahlman, 1979):

> "when we learn that Clyde is an elephant, we want to create a single VC link from CLYDE to TYPICAL-ELEPHANT and let it go at that, but we want the effect of this action to be identical to the effect of actually copying the entire elephant description, with the CLYDE node taking the place of TYPICAL-ELEPHANT. It must be possible to augment or alter the information in this imaginary description without harming the original.

It must be possible to climb around on the imaginary copy-structure and to access any part of it in about the same amount of time (speaking in orders of magnitude) that would be required if the copy had actually been made. But we want all of this for free. We just cannot afford the time or memory-space necessary to actually copy such large structures whenever we want to make an instance or a sub-type of some type-node..."

The notion of compactness through shared structure is pervasive in inheritance systems, but virtual copies goes beyond that due to a feature of NETL known as roles and existence links. A role used this way is an existentially quantified individual; the feature is similar to roles and paraindividuation in KLONE. If we say that elephants have hearts in NETL (*i.e.*, we give the elephant node a heart role) then every instance of elephant will have a corresponding instance that is its heart. A detailed knowledge of elephants would produce many roles for the elephant node, and many relationships between these roles. The power of the virtual copy idea can really be seen when these complex role structures are instantiated seemingly for free for each instance. For example, if the system knows that elephants have lungs and that an elephant's heart is just below his lungs, then when we make Clyde be an instance of elephant the system will automatically know that Clyde has a heart and lungs, and that Clyde's heart is just below Clyde's lungs, not the typical elephant's lungs.

Unfortunately the true meaning of roles in NETL is as clouded today as its IS-A hierarchy inheritance was before this thesis, since roles are dependent upon the IS-A hierarchy but introduce many new sorts of problems. Some of these are marker propagation related (see Fahlman's discussion of the copy-confusion problem), but others, such as the issue of exceptions to role existence, are purely semantic.

The virtual copy idea doesn't make any difference to the semantics of inheritance systems; by definition we should get the same semantics if we make actual copies of inherited roles. Virtual copies is an implementation goal, *i.e.* we want the system to behave as if it made actual copies but without incurring the memory cost. Although the results in chapters 4 and 6 of this thesis show that PMPM's are not adequate for complex inheritance reasoning, some variant of marker propagation might well be useful for implementing virtual copies if we are willing to pay the price of conditioning to preserve correct semantics.

214

8.11 Observations about knowledge representation in general

A rigorous mathematical approach is so unusual in the area of knowledge representation (outside of a few researchers working on nonmonotonic logic, circumscription, and the like) that some justification seems in order. First order logic is by far the most widely used formal representation language, yet much of the knowledge that interests AI researchers cannot at present be conveniently represented in FOL. Consequently, some AI researchers turned their backs on formal systems altogether and developed new, *ad hoc* representations that they believed were better suited to what they have to say.

The promise of these representations is that they introduce powerful new concepts that can make for a much more powerful representation language, as was the case when default reasoning first appeared in frame systems. (Winograd (1980) notes that Micro-Planner's THNOT could also be used as a default reasoning device.) The problem, though, is that if a representation is defined in an *ad hoc* fashion there is no way to say what it really means. Questions about consistency, redundancy of information, and correctness of implementations cannot be answered by purely informal methods.

New formalisms for representing knowledge are usually presented via a set of intuitive examples. Examples are helpful to the reader, but they are no substitute for precise definitions. An example of the limitations of the informal approach to knowledge representation can be found in the history of NETL. Inheritance in NETL was defined by example and supported by appeals to the reader's intuition. Later it was discovered that the algorithms in (Fahlman, 1979) did not implement this intuition correctly (Fahlman, Touretzky, and van Roggen, 1981). More seriously, our intuition turned out to be vague in some areas, and since it had never been formalized, it was not always possible to tell from a NETL diagram what its intended meaning was. When we formulated what we thought was the correct informal theory, it still turned out to have severe flaws only discovered much later. Another problem was that there was no precise characterization of the sorts of computations a parallel marker propagation machine was capable of, and the sorts that require a more powerful architecture than a PMPM, such as a message passing architecture.

In this thesis I took the intuition about inheritance that underlies NETL, refined it, formalized it, and rigorously analyzed the result. A major contribution of the work is that it suggests that some intuitive representations *can* be formalized and treated with

215

mathematical precision, although to do so may require ventures outside the familiar territory of first order logic. The solution that was found once the need for a rigorous approach was accepted might serve as a basis for adaptation to formalizing other representation systems.

References

Allen, J. F. and Frisch, A. M. What's in a semantic network? *Proceedings of the 1982 ACL Conference.*

Altham, J. E. J. *The Logic of Plurality.* Methuen and Company, Ltd., London, (1971).

Anderson, J. R. *The Architecture of Cognition.* Harvard University Press, Cambridge, MA, (1983).

Attardi, G., and Simi, M. Consistency and completeness of Omega, a logic for knowledge representation. *Proceedings of IJCAI-81*, Vancouver, British Columbia, (1981), 504-510.

Attardi, G., and Simi, M. Semantics of inheritance and attributions in the description system Omega. AI Memo No. 642, MIT Artificial Intelligence Laboratory, Cambridge, MA, (1982).

Bobrow, D. G. and Winograd, T. An overview of KRL, a knowledge representation language. *Cognitive Science* 1(1), January (1977), 3-46.

Bobrow, D. G. and Stefik, M. The LOOPS manual (preliminary version). Working paper KB-VLSI-81-13, Knowledge-Based VLSI Design Group, Xerox Palo Alto Research Center, Palo Alto, CA, August, (1981).

Borning, A. H. and Ingalls, D. H. H. Multiple inheritance in Smalltalk-80. *Proceedings of AAAI-82*, Pittsburgh, PA, (1982), 234-237.

Brachman, R. J. "I lied about the trees." *AI Magazine* 6(3):80-93, Fall (1985).

Brachman, R. J. *A Structural Paradigm for Representing Knowledge.* Ablex Publishing Corp., Norwood, NJ, (in press). Original version appeared as technical report 3605, Bolt Beranek, and Newman, Inc., Cambridge, MA, (1978).

Brachman, R. J., Fikes, R. E., and Levesque, H. J. KRYPTON: integrating terminology and assertion. *Proceedings of AAAI-83*, Washington, DC, (1983), 31-35.

Brachman, R. J. and Schmolze, J. G. An overview of the KL-ONE knowledge representation system. *Cognitive Science* 9(2):171-216, (1985).

Carnese, D. J. Multiple inheritance in contemporary programming languages. Doctoral dissertation, TR-328, MIT Laboratory for Computer Science, Cambridge, MA, (1984).

Collins, A. Fragments of a theory of human plausible reasoning. In D. L. Waltz (Ed.) *Theoretical Issues in Natural Language Processing-2*. University of Illinois, Urbana-Champaign, (1978), 194-201.

Dahl, O. J. Simula 67 common base language. Technical report, Norwegian Computing Center, Oslo, (1968).

Department of Defense. *Reference Manual for the Ada Programming Language*. Ada Joint Program Office, Washington, DC, (1982).

Doyle, J. Some theories of reasoned assumptions. Technical report CMU-CS-83-125, Computer Science Department, Carnegie-Mellon University, Pittsburgh, PA, (1983).

Etherington, D. W. and Reiter, R. On inheritance hierarchies with exceptions. *Proceedings of AAAI-83*, Washington, DC, (1983), 104-108.

Etherington, D. W. Formalizing non-monotonic reasoning systems. Technical report 83-1, Department of Computer Science, University of British Columbia, Vancouver, Canada, (1983).

Fahlman, S. E. *NETL: A System for Representing and Using Real-World Knowledge*. The MIT Press, Cambridge, MA, (1979).

Fahlman, S. E. Design sketch for a million-element NETL machine. *Proceedings of AAAI-80*, Stanford, CA, (1980), 249-252.

Fahlman, S. E., Touretzky, D. S., and van Roggen, W. Cancellation in a parallel semantic network. *Proceedings of IJCAI-81*, Vancouver, British Columbia, (1981), 257-263.

Fahlman, S. E. Three flavors of parallelism. *Proceedings of the Canadian Society for Computational Studies of Intelligence-82*, Saskatoon, Canada, (1982), 230-235.

Fox, M. S. On inheritance in knowledge representation. *Proceedings of IJCAI-79*, Tokyo, Japan, (1979), 282-284.

Hayes, P. J. In defence of logic. *Proceedings of IJCAI-77*, Cambridge, MA, (1977), 559-565.

218

Hayes, P. J. The logic of frames. In B. L. Webber and N. J. Nilsson (Eds.) *Readings in Artificial Intelligence.* Tioga Publishing Co., Palo Alto, CA, (1979a), 451-458.

Hayes, P. J. The naive physics manifesto. In D. Michie (Ed.) *Expert Systems in the Micro-Electronic Age.* Edinburgh University Press, Edinburgh, Scotland, (1979b).

Hayes, P. J. and Hendrix, G. G. A logical view of types. *Proc. of the Workshop on Data Abstraction, Databases and Conceptual Modelling.* SIGART Newsletter No. 74, January (1981), 128-130.

Hewitt, C., Attardi, G., and Simi, M. Knowledge embedding in the description system Omega. *Proceedings of AAAI-80*, Stanford, CA, (1980), 157-164.

Hillis, W. D. *The Connection Machine.* The MIT Press, Cambridge, MA, (1985).

Hirst, G. and Charniak, E. Word sense and case slot disambiguation. *Proceedings of AAAI-82*, Pittsburgh, PA, (1982), 95-98.

McCarthy, J. Circumscription – a form of non-monotonic reasoning. *Artificial Intelligence* 13(1,2):27-39, April (1980).

McDermott, D. and Doyle, J. Non-monotonic logic I. *Artificial Intelligence* 13(1,2):41-72, April (1980).

McDonald, D. B. Understanding noun compounds. Doctoral dissertation, Computer Science Department, Carnegie-Mellon University, Pittsburgh, PA, (1982).

MacLane, S. and Birkhoff, G. *Algebra.* The Macmillan Company, New York, (1967).

Minsky, M. and Papert, S. *Perceptrons.* The MIT Press, Cambridge, MA, (1969).

Minsky, M. A framework for representing knowledge. In P. H. Winston (Ed.) *The Psychology of Computer Vision.* McGraw-Hill Publishing Co., New York, (1975).

Moore, R. C. Semantical considerations on nonmonotonic logic. Technical note 284, SRI International, Menlo Park, CA, (1983).

Nilsson, N. J. *Principles of Artificial Intelligence.* Tioga Publishing Co., Palo Alto, CA, (1980).

Parker-Rhodes, F. J. *Inferential Semantics.* Humanities Press, New Jersey, (1978).

Quillian, M. R. Semantic memory. In M. Minsky (Ed.) *Semantic Information Processing*. The MIT Press, Cambridge, MA, (1968).

Reiter, R. A logic for default reasoning. *Artificial Intelligence* 13(1,2):81-132, April (1980).

Reiter, R. and Criscuolo, G. On interacting defaults. *Proceedings of IJCAI-81*, Vancouver, British Columbia, (1981) 270-276.

Roberts, R. B. and Goldstein, I. P. The FRL manual. AI Memo No. 409, MIT Artificial Intelligence Laboratory, Cambridge, MA, (1977).

Touretzky, D. S. Implicit ordering of defaults in inheritance systems. *Proceedings of AAAI-84*, Austin, TX, (1984), 322-325.

Touretzky, D. S. Inheritable relations: a logical extension to inheritance hierarchies. *Proceedings of the Workshop on Theoretical Approaches to Natural Language Understanding*, CSCSI/SCEIO, Halifax, Nova Scotia, (1985), 55-60.

Weinreb, D. and Moon, D. Lisp machine manual. MIT Artificial Intelligence Laboratory, Cambridge, MA, (198).

Winograd, T. Extended inference modes in reasoning by computer systems. *Artificial Intelligence* 13(1,2):5-39, April (1980).

Woods, W. A. Taxonomic lattice structures for situation recognition. In D. L. Waltz (Ed.) *Theoretical Issues in Natural Language Processing-2*. University of Illinois, Urbana-Champaign, (1978), pp. 33-41.

Wright, J. M. and Fox, M. S. SRL/1.5 user manual. Robotics Institute, Carnegie-Mellon University, Pittsburgh, PA, (1983).